Language-driven Exploration and Implementation
of Partially Re-configurable ASIPs

Anupam Chattopadhyay · Rainer Leupers ·
Heinrich Meyr · Gerd Ascheid

Language-driven Exploration and Implementation of Partially Re-configurable ASIPs

Springer

Anupam Chattopadhyay
CoWare India Pvt. Ltd.
2ndFloor, Tower-B
Logix Techno Park,
Sector-127
Noida-201301
India
anupamc@coware.com

Prof. Dr. Heinrich Meyr
RWTH Aachen
Templergraben 55
52056 Aachen
Germany
heinrich.meyr@iss.rwth-aachen.de

Prof. Dr. Rainer Leupers
RWTH Aachen
Templergraben 55
52056 Aachen
Germany
leupers@iss.rwth-aachen.de

Prof. Dr.-Ing. Gerd Ascheid
RWTH Aachen
ISS-611810
Templergraben 55
52056 Aachen
Germany
gerd.ascheid@iss.rwth-aachen.de

ISBN: 978-1-4020-9296-1   e-ISBN: 978-1-4020-9297-8

DOI 10.1007/978-1-4020-9297-8

Library of Congress Control Number: 2008936879

Printed on acid-free paper

9 8 7 6 5 4 3 2 1

springer.com

Acknowledgements

LISA was, is and will be a sweetheart. I mean, the language LISA. From late 2002 till today, it provided a cozy atmosphere for me to keep on imagining and experimenting. I thought of things, which processor designers will be afraid to implement, simply because of its complexity. It was just few tweaks, few keywords in LISA. I am deeply grateful to all those, who thought of ADL at the first place and in particular LISA. Without that, this research would not be possible. I am grateful to the persons who helped me to grasp LISA (thanks to Oliver, David, Manuel, Diandian), to the persons who incessantly questioned me on the abilities of LISA (thanks to Harold, Marcel, Bastian, Maximillian, Benedikt, Diandian, Zoltan) and of course to those who led parallel efforts to extend the language, unknowingly igniting my spirit from time to time (thanks to Tim, Oliver, David, Martin, Torsten, Gunnar, Achim, Kingshuk).

Brainstorming is an important research ingredient. I was fortunate to have co-students, who were always eager to listen to my numerous wacky-tacky ideas and correct me ruthlessly. I especially remember David, Kingshuk and Martin, without whom, I would probably waste experimenting to find out that I was wrong. David is the guy, who helped to design a basic block of this work. He made sure that his component movement is working, even when it was hardly needed for him to do so. Harold worked hard in parallel with me to make sure that the case studies work. Without him, the W-CDMA would never be ready in time. The same goes true about Hanno, who toiled for hours to get the C Compiler up and running, whenever I requested so. The off-topic discussions I had with my co-students at ISS were full of life. It helped me to nurture extravagant ideas all the time. I thank them for their warm association.

At some point of time in the past 5 years, I realized that the sheer breadth of the topic is getting huge. I needed to concentrate on one topic, abandoning the rest ideas as futile dreams. I refused to do so. To this effort, I received enthusiastic support from students, who worked with me towards their theses. I could never have thought of low power and re-configurability together without them. I sincerely offer my gratitude to Martin, Diandian, Benedikt, Zoltan, Waheed, Xiaolin, Ling, Yan, Arnab, Shravan, Eike, Yilin, Ankit and Marcel.

When I offer my suggestions to aspirant PhD students, I remind them repeatedly to choose the Professor and everything else will follow. It could not be truer in my case. I received critical feedback, brilliant insights, lofty goals and freedom to think during my ISS days. That was possible because of the environment created here by Professor Meyr, Professor Ascheid and Professor Leupers. I dared to imagine and refused to give up - mostly because of them. Professor Leupers enriched me by sharing his experiences and by showing strong confidence in my work, Professor Meyr kindled my enthusiasm by his passion for technical innovation and Professor Ascheid offered critical judgement to my work, evidently followed by a definite suggestion. Having a mind-set that research needs to be done in the most austere way, I learned that the path is not less important than the goal. I realized that serious research can co-exist with say, gliding across mountains. I thank them for what I have achieved and learned during last 5 years.

On the personal front, I found many believers throughout my life. Since my early childhood, they embraced me during failure, spiritedly celebrated my success and always bestowed utmost confidence in me. That was essential. I am blessed to have a caring and loving family around me as well as all-weather friends. They kept me going all the time. I wish, someday, I could pass this priceless wealth to the greater world around me.

Contents

Chapter 1
Introduction

> *Nothing endures but change.*
> Herakleitos, Philosopher, circa 535–475 BC

The world is witnessing phenomenal growth in the field of information technology during the last few decades. Throughout the growth, the scientific and consumer market remained the major driving force behind. Chiefly, applications such as wireless communication, network processing, video games ramped up with ever increasing complexity. This complexity is handled simultaneously by two drivers in the system design community. First, there is continuous downscaling of fabrication technology [1] allowing tremendous processing power to be crammed in on a single integrated circuit. Second, there is a design trend of having multi-component systems to address the conflicting requirements of increasingly complex applications. The system components typically include processors, Application-Specific Integrated Circuits (ASICs), Field-Programmable Gate Arrays (FPGAs). In this twin trend of system design, the system-level design upscaling clearly failed to leverage the advantages offered by the current mainstream fabrication technology. This failure, referred as *crisis of complexity* [2], boosted strong research in the field of Electronic System Level (ESL) design.

ESL design automation tools, in varied form, attempt to answer one primary concern. That is, to determine the optimal combination of components for targeting a particular application domain and integrate them. This indicates a strong requirement of tools to explore the huge design space of system components and their integration aspects within a short time-to-market window. The challenge on ESL design tools have grown even more due to the shortening time-to-market window and fast-evolving applications. Furthermore, the high-paced technological growth resulted into skyrocketing mask production cost. In a nutshell, the designers have to find the best-in-class system for the application within a short time with a long time-in-market potential to amortize the design costs. This fuelled an increasing attention towards designing flexible, yet high-performance systems. As can be seen in Fig. 1.1, flexibility is traded off with performance across different system components.

Flexibility in a digital system can be planted in two forms. First, there are the systems which contain *soft* flexibility e.g. processors. However, the degree of soft flexibility in a processor varies. A General Purpose Processor (GPP) is more flexible than an Application-Specific Instruction-set Processor (ASIP). The higher degree

A. Chattopadhyay et al., *Language-driven Exploration and Implementation of Partially Re-configurable ASIPs*, DOI 10.1007/978-1-4020-9297-8_1,

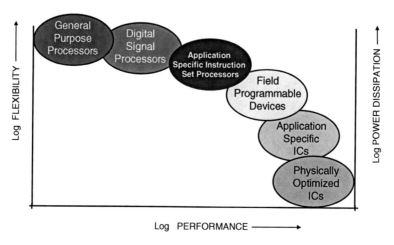

Fig. 1.1 Power-performance-flexibility analysis of system components

of flexibility arises from a more universal nature of the Instruction-Set Architecture (ISA) of the processor. ASIPs employ special instructions and corresponding hard-designed functional units targeted towards a set of applications. Another set of applications may not map well to this application-specific ISA. Contrarily, the GPP keeps on delivering a standard performance over various set of applications. Second, there are other sets of systems offering *hard* flexibility e.g. Programmable Logic Devices (PLD). For these kind of devices, the hardware can be customized by changing the interconnects between various functional blocks. Due to the explicit parallelism offered by PLDs, it delivers high performance in regular data-driven applications. For control-driven applications, PLDs cost high power overhead due to the long running interconnects over the complete device. In order to reap the benefits of both kinds of flexibility, a new architecture class have emerged over the years. This architecture contains a fixed processor part coupled with a re-configurable block. We refer to these architectures as re-configurable ASIPs (rASIPs). The re-configurable block can be statically or dynamically re-configurable, depending on which, the architecture is called s-rASIP or d-rASIP respectively. The degree of flexibility offered by a rASIP can be tuned at various levels. By having the base processor, soft flexibility is already available in form of the ISA. Further soft flexibility can be introduced by reserving some instructions to be locally decoded and executed by the re-configurable block. The resulting instruction-set of the rASIP is *morphable* i.e. can be extended to cover a new application domain. Finally, the customizable hardware architecture in form of PLDs provides rASIPs with hard flexibility.

From the perspective of design productivity, rASIPs threw up a new challenge to the processor design community. The complete processor, re-configurable block and the interfacing must be designed to suit a range of applications. Figure 1.2 shows several alternative rASIP designs. The re-configurable block, as shown in the

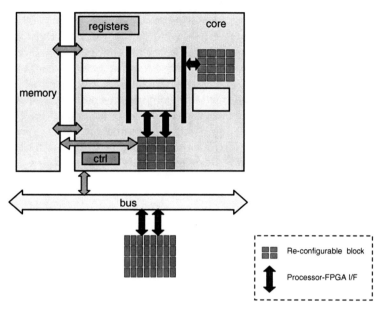

Fig. 1.2 Alternative rASIP designs

figure, can connect to the base processor in a loosely coupled manner i.e. through an external bus. It can be tightly coupled to the base processor, where it acts like an internal functional unit of the processor. Leaving apart the position and selection of the re-configurable block, the base processor architecture (RISC, VLIW, SIMD) and the FPGA architecture of the re-configurable block presents with numerous design choices. The design space offered by rASIP is huge, of which the optimum point must be determined within a competitive time-to-market window.

This book presents a solution to this design complexity problem by providing a workbench for rASIP design space exploration. A high-level language is developed for modelling the rASIP architecture. The necessary tools for pre-fabrication and post-fabrication rASIP design space exploration are either derived from the high-level language automatically or are generically developed standalone. Several case studies show the feasibility of this language-based rASIP design approach. Furthermore, the framework opens a new horizon of possibilities by allowing to model partially re-configurable processors using language-based abstraction. This approach not only benefits rASIP design, but also can be applied to modelling of PLDs standalone. Given the capacity of the proposed language to seamlessly model a range of components, it can very well be pitched as a candidate for modelling all varieties of programmable systems.

This book is organized as following. The following chapter, Chapter 2 provides the background of processor design and the development of processor description languages. This chapter chronologically outlines the development of processor description formalisms. Chapter 3 describes the related work in the field of partially re-configurable processor design methodology and sets the motivation of this book. The advantages and drawbacks of various approaches are also clearly mentioned. The rASIP design space is elaborated in Chapter 4 followed by an overview of the proposed rASIP design flow. Chapters 5 and 6 concentrates on pre-fabrication design space exploration and design implementation for rASIPs, whereas Chapter 7 gives detailed account of the tools and methodologies for post-fabrication design space exploration and design implementation respectively. Several case studies, which uses the proposed methodology of this book, partially or fully, are performed during the development of this work. Those case studies are detailed in Chapter 8. The summary and outlook of this book are presented in Chapter 9.

Chapter 2
Background

Education is not the filling of a pail, but the lighting of a fire.
William Butler Yeats, Poet, 1865–1939

2.1 Processor Design : A Retrospection

To develop computers which are going to be software compatible, IBM originally designed System/360, the first commercially available Complex Instruction Set Computing (CISC) processor. In the process of achieving software programmability, System/360 also made a clear separation between *architecture* and *implementation* [3]. There the architecture was described as the attribute of the system as seen by the programmer. The logical design, organization of the data flow and controls were classified as implementation. CISC processors typically contained a complex instruction-set i.e. an instruction capable of performing a series of sequential and/or parallel operations on a set of data. During the mid-1970s, several research projects demonstrated that the processors with numerous complex instructions are overdesigned for many applications. Another fallback of the CISC processors was that the high-level language compiler was inefficient compared to a manual assembly programmer. This led to underusage of many features offered by CISC. Consequently, the philosophy of Reduced Instruction Set Computing (RISC) dawned with several parallel research endeavors [4, 5]. The first processor developed at Berkeley RISC project, known as RISC-I, outperformed the other single chip design at that moment with only 32 instructions [6]. The small set of instructions, typical of RISC paradigm, encompassed a wider range of applications without wasting hardware area. Clearly it was more flexible and easier to adapt than CISC processors. With the advent of ubiquitous computing during 1990s, a demand of high-performance (to meet the application complexity) yet flexible (to be designed and retargeted for a wide domain of applications) processors rose. RISC processors fitted the bill perfectly. As a result, RISC philosophy were widely adopted for embedded microprocessors.

However, it was noted that, the flexibility coming from the programmability of the processor sacrifices performance and power. Therefore, efforts were made to balance between the flexibility over a range of applications and performance. This resulted into the interesting swing of system design community among customization and standardization, leading to the postulation of Makimoto's Wave [7] (see Fig. 2.1). According to this wave, the design community swings between

A. Chattopadhyay et al., *Language-driven Exploration and Implementation of Partially Re-configurable ASIPs*, DOI 10.1007/978-1-4020-9297-8_2,
© Springer Science+Business Media B.V. 2009

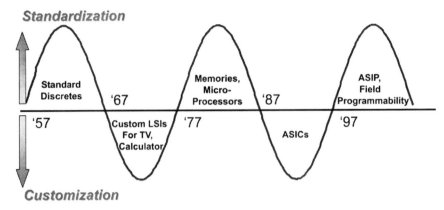

Standardization

Standard
Discretes

'67

'57

Custom LSIs
For TV,
Calculator

'77

Memories,
Micro-
Processors

'87

ASICs

ASIP,
Field
Programmability

'97

Customization

Fig. 2.1 Makimoto's wave

standardization and customization of system components. Needless to say, this wave is visible within the processor design arena as well. With customization the processor instruction set is derived by studying the application, thereby presenting the class of Application-Specific Instruction-set Processors (ASIPs) [8]. ASIPs can be classified as a resurgence of CISC-like instructions within the RISC paradigm. With standardization, a wider range of applications are targeted by designing domain-specific processors [9]. With customization, a narrow domain of application is targeted with highly specific computation engines. Understandably, the embedded processor design community required an ever stronger support from the design automation tools to explore the choices and rightly settle the flexibility-performance trade-off.

The design of a processor is a highly involved task requiring diverse skills. Once the target group of applications is selected, an Instruction Set Architecture (ISA) and the micro-architecture needs to be designed. Furthermore the tools for mapping the application, typically written in a high-level language, to the ISA and the micro-architecture needs to be designed. As observed in case of the early CISC processors, a poor mapping tool can drastically undermine the true performance potential of the processor. Since the design of software tools in contrast with the hardware micro-architecture requires different skill-set, traditionally the processor design is performed by two separate groups of designers. First, one group does the ISA design, software tool development and then the other group works on the micro-architecture modelling as per the ISA specification. This kind of sequential design approach is time-consuming. It is also difficult to account for late design changes. To get rid of this, parallelization of the software development and hardware development for the processor is advocated. Even then, the following issues remain.

1. *Correct Functionality* : The interface between these two groups need to be uniquely determined throughout the complete design and verification. Lack of proper synchronization between the two groups may lead to functional error or long re-design cycles. The interface can be a high-level specification. For correct

functionality and early validation, it is important that this specification is executable (at least on a host processor).

2. *Optimality :* Design changes in the ISA and/or in the hardware micro-architecture need to be taken into account by the other group to ensure optimal performance. For example, the handling of data-hazards may be covered by the hardware dataflow organization by having data forwarding paths. The C-compiler must be aware of that in order to do the optimal scheduling.

These issues become even more challenging due to the large number of design options. The alternative design decisions need to be weighed against each other to reach the optimal point in the entire design space. With a slow and error-prone design methodology, this is hardly possible.

2.2   High-level Processor Modelling

To reduce the complexity of processor design, researchers proposed abstract modelling of the entire processor. The abstract processor model, in effect can serve as an executable specification for early design space exploration. The major example of this abstract processor modelling is found in Architecture Description Languages (ADLs) [10, 11, 12, 13, 14]. Using an ADL, the entire processor can be described. To hold both the ISA and the micro-architecture in the same description, the ADL allows constructs such as assembly syntax along with data-flow description. From the ADL description, software tools e.g. C-compiler, instruction-set simulator as well as the hardware micro-architecture can be automatically or semi-automatically generated. With the above characteristics in common, the ADLs differ in their focus, efficiency and modelling style. A brief overview of prominent ADLs are given in the following paragraphs. A detailed discussion can be found in [15].

nML   nML architecture description language is based on the concept of hierarchical instruction-set organization. The instructions with common dataflow are grouped and placed higher in the organization. The special dataflow is placed in the elements lower in the hierarchy and termed as Partial Instruction (PI). The partial instructions can be grouped in their parents by two kinds of rule e.g. AND-rule and OR-rule. The OR-rule specifies two or more alternative PIs corresponding to an instruction. The AND-rule, which can be existent orthogonally to the OR-rule, specifies the composition of parts of an instruction. A complete instruction is formed by traversing down the tree following a particular branch. The following code snippet (refer to Fig. 2.2) helps to understand the concept.

   Here, the two PIs `loadoperate` and `control` are connected via the OR-rule within the topmost instruction indicating that the instruction can be of either of these two types. Within PI `loadoperate`, PI `load` and PI `operate` are again connected via the AND-rule. This means that the PI `loadoperate` is composed of the two PIs `load` and `operate`. Each PI consists of several

```
opn instruction = loadoperate | control

opn loadoperate(l:load, o:operate)
    image  = "010"::l.image::o.image
    syntax = format("\%s || \%s", l.syntax, o.syntax)
    action =
    {
        l.action();
        o.action();
    }
```

Fig. 2.2 Part of an nML description

sections for describing the instruction coding (*image* section), assembly syntax (*syntax* section) and the execution behavior (*action* section). While generating a complete instruction by traversing through the tree, these sections can be combined to form the overall instruction syntax or coding. To model the structural organization of the underlying processor micro-architecture, nML uses two kinds of storage elements. Firstly, there are *static* storage elements, which hold the value until it is overwritten. These are used to model memories and registers in the processor. Secondly, there are *transitory* storage elements, which hold the value for a given number of cycles only. These storages can be explicitly synchronized to model the effect of pipeline registers, thereby implicitly dividing the entire dataflow into pipeline stages.

nML is one of the first ADLs to introduce hierarchical instruction-set organization, combining ISA and execution behavior in the same model. These novel features allowed easy generation of the instruction-set simulator and retargetable code generator for the compiler. From nML, software tools and RTL description of the processor can be automatically generated. nML can also be used to generate test patterns for early processor verification. Currently, nML and its associated tools are commercially available [16].

LISA In LISA, an *operation* is the central element to describe the timing and the behavior of a processor instruction. The instruction may be split among several LISA operations. Similar to nML, the LISA description is based on the principle that the common behavior is described in a single operation whereas the specialized behavior is implemented in its *child* operations. With this principle, LISA operations are basically organized as an n-ary tree. However, specialized operations may be referred to by more than one *parent* operation. The complete structure is a Directed Acyclic Graph (DAG) $\mathcal{D} = \langle V, E \rangle$. V represents the set of LISA operations, E the graph edges as set of child-parent relations. These relations represent either *Behavior Calls* or *Activations*, which refer to the execution of another LISA operation. Figure 2.3(a) gives an example of a LISA operation DAG. A chain of operations, forming a complete branch of the LISA operation DAG, represents an instruction in the modelled processor. In a modelling extension over the existing nML description, LISA allowed explicit pipeline stage assignment for the operations. Each operation

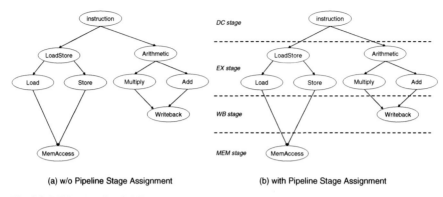

(a) w/o Pipeline Stage Assignment   (b) with Pipeline Stage Assignment

Fig. 2.3 LISA operation DAG

can be explicitly assigned to a pipeline stage. The corresponding LISA Operation DAG looks like as shown in the Fig. 2.3(b).

The processor resources (e.g. registers, memories, pipeline registers) are declared globally in the *resource section* of a LISA model, which can be accessed from the operations. Registers and memories behaved like static storage in contrast with the transitory storing nature of pipeline registers. Apart from the resource declarations, the resource section of LISA is used to describe the structure of the micro-architecture. For this purpose the keywords *pipeline* and *unit* are used. The keyword *pipeline* defines the instruction pipeline of the processor with the corresponding order and name of the pipeline stages. The *pipeline_register* is used to define the pipeline registers between the stages. Using the keyword *unit*, the designer can define a set of LISA operations (within a pipeline stage) to form an entity (VHDL) or module (Verilog) in the generated HDL code. LISA allowed one further designer-friendly feature by allowing arbitrary C-code to model the LISA operations' execution behavior. A more detailed description of the LISA language elements is provided in 4.

LISA has been used to generate production-quality software tools, C-compiler and optimized HDL description. The language and its associated tools are currently available via [17].

EXPRESSION   An EXPRESSION description consists of two major sections namely, *behavior* and *structure* reflecting the separation between the ISA and the micro-architecture. The behavior section comprises of three sub-sections. Firstly, there is an *operation* sub-section. Within the operation sub-section, the opcode, operands and the execution behavior of the operation is specified. Secondly, there is *instruction* sub-section. Each instruction contains slots which contain *operations* to be executed in parallel. Each of these slots again correspond to a functional unit in the micro-architecture. Finally, there is an *operation mapping* sub-section for specifying information needed for code selection and performing architecture-dependent compiler optimization.

The structure section of the EXPRESSION language is again composed of three sub-sections. These are *component*, *memory subsystem* and *pipeline/*

data-transfer path. The component sub-section describes functional units, pipelined multi-cycle units, ports and connections. The memory subsystem provides the description of different storage components such as SRAM, DRAM, cache with their detailed access information. The pipeline/data-transfer path sub-section describes the overall processor in a netlist representation, albeit in a coarse-grained manner. The pipeline path specifies the units of which a pipeline stage is composed. A data-transfer path specifies all valid data-transfers between the elements of pipeline, component and memory subsystem. In the following Fig. 2.4, a processor architecture is shown indicating different elements of EXPRESSION language.

In contrast to LISA and nML, EXPRESSION offered more rigor in the language organization. In EXPRESSION, the instructions are modelled in the structural part of the language. As an outcome, the freedom of modelling arbitrary execution behavior is more limited in case of EXPRESSION. On the flip side, due to strong emphasis on structures, resource sharing between instructions, extraction of reservation tables for superscalar architectures become easier for EXPRESSION.

The language and toolkit for EXPRESSION is available as an open source release in the public domain [18].

All the ADLs encourage the processor designer to take a workbench approach. The ADL can be used to describe the initial version of the processor. The associated tool suite can be generated automatically or semi-automatically. These tools help to map the high-level application on to the processor. A high-level simulation of

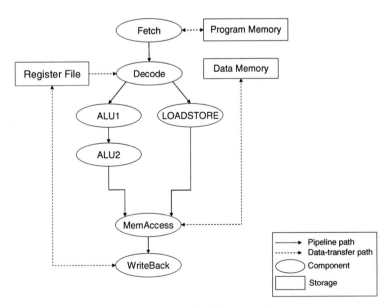

Fig. 2.4 A processor architecture from EXPRESSION's viewpoint

the mapped application reveals the scopes of optimization both in the application and in the architecture. Once the processor description is finalized, the hardware generation from the ADL enables to proceed towards lower level of abstraction. Depending on the offered modelling style, the ADLs are typically classified in structural, behavioral or mixed classes. The more emphasis an ADL puts on behavioral aspects, it becomes easier for the designer to explore design options in early stage. The same ADLs typically find it difficult to generate efficient hardware or make several implicit assumptions during micro-architecture derivation phase. Alternatively, the more structurally rigorous an ADL is, the less freedom it offers to describe highly irregular and wayward architectures. These ADLs, on the other hand, offer advantages in the phase of hardware generation. Independent of these characteristics, for all ADLs, the generation of efficient High Level Language (HLL) compiler turned out to be a major challenge. Usually, the ISA is sufficient for the code generation part of the compiler. However, several micro-architectural aspects need to be taken into account for efficient code generation. For instruction scheduling and register allocation, especially for pipelined micro-architectures, more detailed information is required. This challenge is addressed in ADLs by introducing separate sections in the description. For example, the *operation-mapping* sub-section carries these information for EXPRESSION ADL. For LISA, *semantic* section is introduced for capturing the instruction semantics [19] during pipelined execution. To increase the efficiency of the compiler, further manual tuning via GUI is also suggested [20].

A particularly strong critique against abstract processor modelling approaches is its long design cycle, which is kind of paradoxical. The abstraction is introduced at first place to reduce the complexity, thereby reducing the design cycle. However, the abstraction needed to be passed to lower levels of detailing and each such elaboration calls for another level of verification. In absence of any pre-verified macro, this task is surely demanding. The key of the paradox is the fact that, the abstract processor models encompasses a huge number of design alternatives, exploration of which in itself introduces a new task in the processor design. Several worthy attempts are made to address the verification bottleneck in ADL-driven processor design. Property-driven verification and symbolic simulation are integrated with ADL EXPRESSION [21, 22]. From LISA, automatic test generation and assertion generation are supported [23, 24]. Coverage-driven test pattern generation and full-coverage test pattern generation are also integrated in various ADLs [25, 26]. Even after these efforts, new design is considered more and more risky in presence of skyrocketing fabrication costs. As a result, ADL-based processor modelling is often pitted only as a high-level exploration methodology.

2.3 Library-Based Processor Design

Library-based processor design (popularly referred as configurable processor design) [27, 28] addresses the verification challenge directly. In this approach, the components of processor are available as pre-verified. The designer can select

among these components and plug those to design the complete processor. Given
a number of alternative components are present, the various combinations for the
complete processor is limited and thus, can be verified completely. Therefore, the
library-based design approach can, up to a certain extent, guarantee the correct functionality of the complete processor for any configuration.

ARC ARC [28] commercially licenses configurable cores for digital signal processing. It offers a basic set of RISC-based processor families, which can be
extended to meet various constraints. The extensions include floating point operations, advanced memory subsystem with address generation, special instruction
extensions (e.g. viterbi, 24 × 24 MAC). A GUI-based configuration tool, named
ARChitecht, allows the designer to select from the pre-configured options and/or
create custom instruction extensions. ARC configurable processors allow mixing
between 16-bit and 32-bit instructions without any switching overhead to optimize code density.

Tensilica Another leading example of library-based processor design approach is
Tensilica [29]. A Tensilica core named, Xtensa 7 is shown in the following Fig.
2.5.

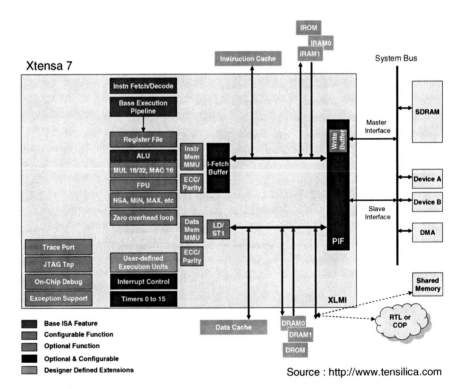

Fig. 2.5 A Tensilica configurable processor architecture

As can be observed from the figure, a large number of configurable or designer-defined extensions can be plugged in to the basic processor core. Custom instructions for application-specific runtime speed-up can be plugged in using Tensilica Instruction Extension (TIE) language. Additionally, Tensilica offers a tool-suite for automatically identifying hot-spots in an application and correspondingly deriving the custom instruction. The complete tool-flow allows fast design space exploration down to the processor implementation. For every possible processor configuration, software tools e.g. compiler, simulator as well as the synthesizable RTL for the micro-architecture can be generated. Verification of various processor configurations is much more easily tackled than ADL-driven processor design. For most of the processor configurations, correct functionality is guaranteed [30]. Tensilica cores demonstrated high performance over a wide variety of contemporary processors.

By having a primary core and additional configurations, Tensilica processors attempted to partition the processor design space. With such coarse partitioning it is difficult to conceive irregular architectures suited for several application domains. Several primary processor settings e.g. memory interface, register file architecture can also turn out to be redundant or sub-optimal.

Conceptually, the library-based processor modelling can be considered as a natural course of path adopted earlier by gate-level synthesis. In gate-level synthesis, the RTL structure is mapped to existing libraries containing pre-verified logic cells. Gate-level synthesis also do not allow modelling of arbitrary transistor circuit, unless the circuit is present in the library of cells. Clearly, the physical optimization is sacrificed for library-based modelling to cope with the huge complexity of transistor-level design. However, the optimization of RTL structure is done effectively. The optimization goals like area, timing and power minimization are mapped onto definite mathematical formulations [31, 32]. The problems are solved by employing algorithms known to deliver optimal or near-optimal results.

The same kind of optimization approach is hardly feasible when applied to processor synthesis. Unlike a fixed hardware block, the processor retains its flexibility through ISA after fabrication. Therefore, the processor designer must chose between flexibility and performance. It is non-trivial to map this goal to a mathematical formulation because of the difficulty to capture flexibility mathematically. This directed the researchers to adopt intuitive and fast design space exploration as the prime mean of processor design. Library-based modelling allowed to reduce the verification effort, while sacrificing the design space. In contrast, ADL-based processor design allowed the wider design space at the cost of increased verification effort.

2.4 Partially Re-configurable Processors : A Design Alternative

Partially re-configurable processors contain a re-configurable hardware block along with the base processor, thereby enhancing the post-fabrication flexibility of processors. Although there is a significant research and commercial effort pouring on partially re-configurable processors in recent times [33, 34], the idea of combining these two are not so new [35, 36]. The early attempts, as we will see in the next chapter, met with the huge challenge of primitive processor design tools and lack of sophisticated re-configurable block architectures.

Any processor remains flexible after the fabrication phase via the software programmability. This kind of flexibility, referred henceforth as *soft* flexibility, is the key to wide design re-use and compatibility across applications [3]. Field-programmable hardware architectures [37, 38] provide another kind of *hard* flexibility by allowing the designer to alter the hardware configuration after fabrication. Field-programmable hardware blocks can be defined as array of fine/coarse-grained functional units, the functionality and interconnection of which can be programmed after fabrication. Commercially available Field-Programmable Gate Arrays (FPGAs) [39, 40] and Field-Programmable Object Arrays (FPOAs) [41] are major examples of such programmable hardware block. FPGAs are traditionally limited to the role of prototyping hardware circuits. The huge speed advantage provided by FPGA compared to software simulation of hardware blocks justified such approach. However, with increasing interest in flexible solutions, FPGAs are making their way to the final implementation [33].

By containing the field-programmable hardware, partially re-configurable processors increase its post-fabrication flexibility drastically. The merger of hard and soft flexibility in one processor look promising with the following advantages over traditional non-reconfigurable processors.

- **Leveraging FPGA Computing Power:** When targeted for the massively parallel computing jobs, FPGAs are demonstrated to outperform the high-end microprocessors designed for the same tasks [42]. The raw computing power over the same area is much higher in case of a field-programmable hardware. In field-programmable hardware, the functional units are connected by the designer-directed configuration. Once the connection is established, the state of each computation is passed to the next functional unit, thereby reducing the effort of temporary storage access and instruction decoding, which are usually prevalent in a processor. In case of processors, the overall processing state is stored in a compact manner via instruction encoding. Therefore, processors turn out to be advantageous in case of infrequent, irregular computation. Due to this complementary nature of computing advantage, a field-programmable hardware tied with a base processor delivers a high-performing coverage over wide application range.

  It is interesting to compare this regular-irregular combination of partially re-configurable processors with the macro-world of SoC. In the modern [43] and future [44] SoCs, few general purpose processors are existent to perform the control-intensive functions, whereas a large array of Data Processing Elements (DPE) are present to manage the data-intensive, regular functions. The Fig. 2.6 shows the similar nature of architectural organization done in SoC as in typical partially re-configurable processors. What is more important in both of the

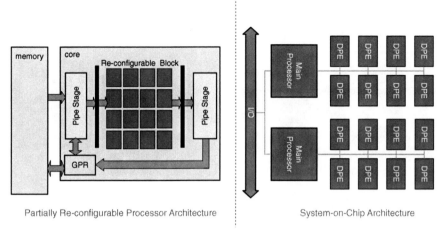

Partially Re-configurable Processor Architecture   System-on-Chip Architecture

Fig. 2.6 Comparison of architecture organization styles : micro and macro

pictures is, that both of these designs are scalable for catering to the increasing design complexity.

- **Increasing Design Re-use:** According to the System-On-Chip (SoC) productivity trend charted by International Technology Roadmap for Semiconductors (ITRS) [45], 38% of the overall design is currently re-used. This percentage is predicted to rise steadily up to 78% in ten years from now. This design re-use is required to maintain the current productivity trend. The system complexity and size becomes prohibitive towards a new design. Re-configurability in hardware adds to the overall system flexibility. The flexible system can be used to fulfill new tasks without being re-designed and without sacrificing required performance. Clearly, partially re-configurable processors promotes design re-use keeping in tandem with the SoC productivity trend.
- **Reduced Verification Risk:** The complexity of modern systems makes it impossible to detect all possible bugs before the system is fabricated. With even pre-verified components, new faults may trigger due to hitherto untested external interfacing. This makes it often imperative to fix design errors after the system is fabricated. Partially re-configurable processors enable designers to perform exactly this via *soft* changes in the program code or via *hard* changes in the field-programmable hardware.

The advantages provided by partially re-configurable processors prompted designers to experiment with this class of architectures [36, 46, 34, 33]. Partially re-configurable processor architectures can be broadly classified in two classes on the basis of the coupling between the processor and the re-configurable hardware. The classification was originally suggested at [47] and later adopted by [48] for a survey on design of partially re-configurable processors. A slightly more detailed classification is suggested in [49]. In the following Fig. 2.7, these organizations are shown.

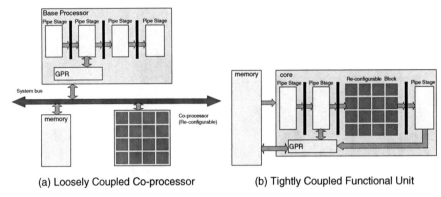

(a) Loosely Coupled Co-processor   (b) Tightly Coupled Functional Unit

Fig. 2.7 Various coupling organizations of partially re-configurable processors

For the loosely coupled co-processor variant, the re-configurable hardware block is placed next to the base processor or next to the memory architecture of the base processor. This kind of arrangement is justified when little communication between the base processor and the re-configurable hardware is required. Prominent examples of these class of partially re-configurable processors are [50, 69]. The architectural arrangement shown in the Fig. 2.7(b) is useful when the amount of communication is much higher. With tight coupling, the re-configurable block is also allowed to access several architectural components, which are not visible in the ISA (e.g. pipeline registers). Here, the re-configurable block can be used here to model custom instructions giving rise to a *morphable* [51] or *dynamic* [52] instruction set, thereby adding a degree of customizability to the ISA.

Due to the broader horizon of architectural possibilities in tightly coupled partially re-configurable processors, this class remains in the primary focus of current book.

2.5 Synopsis

- To increase flexibility without sacrificing performance, various processor micro-architectures and ISAs are experimented with.
- For quick design space exploration among these alternatives, abstract processor modelling and library-based processor modelling design methodologies are proposed.
- With post-fabrication customizability, field-programmable hardware is currently viewed as final implementation choice instead of being used as a prototype.
- Partially re-configurable processors emerge as a new class of processors to combine the merits of irregular control-intensive processor with regular data-intensive FPGA.
- Primary focus of the current book is on tightly coupled partially re-configurable ASIP, referred as rASIP henceforth.

Chapter 3
Related Work

In this chapter, the related work in the field of rASIP design is traced. Initially, various design points are identified by chronologically studying the evolution of partially re-configurable processors. Following that, a brief overview of separately evolving design tools is presented. Finally, the most general approaches of design space exploration and implementation are elaborated. To analyze the existing work in the field of rASIP design, separate emphasis is made on *how* (to understand the design flow) and *what* (to understand the design point) is designed. To appreciate the design methodology, the following items are reviewed as available in the published literature.

- ISA Modelling, Exploration and Implementation

  1. Representation of ISA.
  2. ISA design alternatives.
  3. Evaluation of ISA alternatives.
  4. Generation of software tools for design space exploration.

- Micro-architecture Modelling, Exploration and Implementation

  1. Representation of the micro-architecture.
  2. Design alternatives of the base processor micro-architecture.
  3. Design alternatives of the re-configurable hardware block.
  4. Design alternatives for the rASIP coupling.
  5. Evaluation of the design alternatives.
  6. Mapping of the micro-architecture to physical implementation.

- Interaction of the ISA and the micro-architecture

3.1 A Chronological Overview of the Expanding rASIP Design Space

The traditional processor design approach is adopted for most of the early rASIPs. In this approach, the processor software tools and the micro-architecture implementation are developed in a decoupled manner. As the design points kept on increasing,

A. Chattopadhyay et al., *Language-driven Exploration and Implementation of Partially Re-configurable ASIPs*, DOI 10.1007/978-1-4020-9297-8_3,
© Springer Science+Business Media B.V. 2009

the approach of design space exploration, re-targetable tool-flow and generic design methodologies appeared.

Splash-2 Splash-2 [36, 53] was designed in an era, when processor design itself was a quite demanding task. Including the re-configurable block during the ISA design phase would make the process prohibitive to manage. Splash-2 is conceived as an attached special-purpose parallel processor aiding the host Sparcstation [54] processor. No ISA modelling or modification is considered. The micro-architecture of the splash-2 host is fixed beforehand. Although no public literature is available showing design space exploration for the micro-architecture and ISA of splash-2, yet the design reflects several features allowing flexibility in the execution. The re-configurable hardware block is developed to exploit spatial or temporal parallelism in the application. To access large data streams from the host processor, high I/O bandwidth is allowed. The interface board allows fast data access via DMA from the host. It also contains FIFO storages to decouple the re-configurable block from the host. The computing elements in the re-configurable block are arranged in a linear array. Furthermore, they are fully connected by a crossbar switch. The host processor can access to the memories of the computing elements, set up the DMA channels, receive interrupts from the DMA channels and the computing elements. In that manner, the host processor controls the re-configurable block closely.

PRISM-I, PRISM-II Several novel ideas about rASIP designing were introduced with PRISM-I [35]. In PRISM-I, it is noted that an adaptive micro-architecture can not be designed by the high-level programmer. This is simply because of his/her expertise either in software or in hardware programming. This task is imposed on the High Level Language (HLL) compiler by PRISM-I team. Thus, the challenge is to build a compiler capable of doing hardware-software partitioning, generating software instructions and the corresponding hardware implementation. This idea, termed as *instruction-set metamorphosis*, enable the processor to retain its general-purpose nature while being application-specific when it is demanded. The task of hardware-software partitioning in PRISM architectures is accomplished by *configuration compiler*. Configuration compiler is responsible for identifying function hot-spots (done with manual interaction), generating access functions to replace those and creating a hardware image to be mapped to re-configurable platform. The system architecture consists of a host processor coupled with a re-configurable block. Neither the host processor, nor the re-configurable block are designed for PRISM-I via design space exploration. PRISM-II [55] offer more advanced version of configuration compiler, where a larger subset of C programming language is supported for hardware synthesis. It is also observed that a tighter integration between the host processor and the re-configurable hardware will reduce the data movement penalty. This problem is addressed by designing a system architecture with efficient communication mechanism.

PADDI, PADDI-2 In a notable effort to drive the architecture design by identifying the application domain, the processor PADDI was conceived [56, 57]. PADDI specifically targeted digital signal processing algorithms. It is noted in [56] that, contemporary processors fail to match with the increasing demand of DSP algorithms for intensive computation and data bandwidth. It also identified the limited ability of the existing FPGAs to perform coarse-grained, arithmetic computations in an efficient way. The PADDI micro-architecture contains 4 clusters, each of which is again a collection of 8 execution units (EXUs) connected via a crossbar switch. These EXUs contain dedicated hardware for fast arithmetic computation. Furthermore, the EXUs are equipped with local controller. This allowed a variety of operations to be mapped onto one EXU. In this regard, PADDI marked one of the earliest examples of a multi-FPGA chip with coarse granularity and localized control. PADDI offered an advanced programming environment with a automated compilation from high-level language, an assembler and a GUI-based simulator. The simulator offered debugging environment for the complete PADDI chip with the ability to single-step through execution and modify registers and instructions in runtime.

Spyder Spyder [58] proposed an architecture with several features for the first time in partially re-configurable processor design. It identified the bottlenecks of existing general purpose computing as having a fixed ISA, which fails to model sequence of operators. Furthermore, the existing processors did not allow data-processing with arbitrary data-width. Contemporary FPGAs did solve exactly these issues. The earlier works on partially re-configurable processor attempted to integrate FPGAs in the processor die, mostly focussing on the processor design. On the other hand, the significance of Spyder was to define an easy programming model for high-level programmer.

The programming model of Spyder consists of three well-separated parts e.g. an application program running on the base processor with special subroutine calls for re-configurable block, a file containing the Spyder microcode and the configuration file for the re-configurable hardware. The programmer needs to write these three parts separately in a language subset of C++. The C++-subset description can be directly compiled to a format compatible to FPGA tools. Easy debugging of the complete application program is also allowed by simulating the high-level re-configurable hardware descriptions. The micro-architecture of Spyder is based on pipelined harvard architecture. The memory bandwidth problem for multiple re-configurable execution units running in parallel is addressed by Spyder. To solve this issue, registers are banked and shared among the re-configurable units. To take advantage of the parallelism offered by re-configurable hardware, Very Large Instruction Word (VLIW) model of instructions is used.

Nano Processor The earlier partially re-configurable processors combined a general purpose processor with a re-configurable block. The idea was to retain the general purpose nature with application-wise performance enhancement. Though the PRISM concept is termed as the metamorphosis of instructions, yet in essence it was a function call to the attached processor with special directives. Nano Processor [59] made one of the early attempts to consider the ISA of the processor in a wholesome way.

The core processor is defined to have 6 essential instructions. Around this, new custom assembly instructions can be developed. For each custom instruction, assembly syntax, opcode and the instruction length need to be specified. The flexible assembler, named as *nano assembler*, takes the custom instruction definitions to generate program files. The advantage of these custom instruction is obvious. The instruction-set remained completely flexible like previous design propositions. Additionally, the custom instructions had access to the processor internal registers allowing tight coupling. The complete organization reflects a processor designed for re-targeting up to the assembly level. The custom instruction behaviors are defined using standard synthesis and schematic tools. The execution behavior of the custom instruction are to be verified beforehand using those in an assembly program. To limit the design space within manageable size, several limitations are imposed on the micro-architecture organization. For example, the sub-tasks of a custom instruction are mapped on well-defined pipeline stages.

DISC A conceptual derivative of Nano Processor, DISC [52] advanced the rASIP design space by taking several technological advancements of re-configurable hardware into account. It is noted at [52] that the number of custom instructions is limited by the hardware space. At the same time, run-time re-configuration of the hardware block was becoming a promising approach. With the ability to dynamically re-configure the entire hardware block, the Field-Programmable Gate Arrays (FPGAs) were able to deliver a higher computation density. DISC leverage this feature to improve the overall performance. Even with the slow re-configuration available at that time, it demonstrates reasonable performance improvement. To minimize the recurring partial re-configuration of the FPGA, the custom instruction modules are arranged in a specific order in the FPGA.

PRISC Programmable RISC, abbreviated as PRISC [60] marked another milestone in the evolution of rASIP architectures. During the design of PRISC, it was correctly observed that the data communication between the programmable hardware and the processor is slow in earlier board-level interfaced rASIPs [36]. To ameliorate this shortcoming, the custom functional unit is entirely placed on the processor chip. However, the rest of the processor is not built with re-configurable fabric like in [52] or in [59]. In this concept, PRISC is derived from PRISM [35]. In contrast to PRISM, PRISC allow fine-grained ISA extensions. In PRISC, the custom instructions are modelled via a specific instruction format. In this format, the input–output operands and the configuring information for executing that particular instruction is stored. If the current configured hardware matches with the demanded configuration, then the instruction is executed right away. Alternatively, the hardware block needs to be re-configured before the custom instruction executes. All configuration information for an application are loaded during application compilation to the data segment of application's object file. It was also noted that recurring partial re-configuration is cost-prohibitive. The ISA extensions from the application were identified in a manner to minimize this. Once the candidate custom instructions are selected, hardware implementation is done from the custom instructions' Control Data Flow Graph (CDFG). The PRISC micro-architecture places the custom functional unit in parallel with the regular functional units, thereby allowing a tight coupling with the base processor. The re-configurable implementation of functional unit is inherently slower compared to the dedicated hardware implementation of the same. With parallel processing of regular data-path units and the custom functional unit, there is an evident issue of latency mismatch. The balancing of latency was avoided by earlier rASIP designs, where the overall system clock was slowed down. PRISC tried to limit the logic levels of the custom functional units. By this it is ensured that the latency of a custom instruction would fit within 1 cycle of the contemporary high-speed microprocessors. On the other hand, this limited the possible number of custom instructions.

OneChip, OneChip-98 In the design of OneChip [61] processor, two key hurdles against achieving high-performance rASIP were identified. Firstly, the inflexible co-processor access protocol. Secondly, the low bandwidth of co-processor access. Interestingly, flexible interfacing was reasoned citing embedded controllers as example. This reflects that still partially re-configurable processors were viewed as an advancement of GPP computing. It also shows that the contemporary embedded computing demanded more flexibility out of their systems. OneChip attempted to couple the re-configurable hardware block with a MIPS core in a more close nature than the earlier designs. The re-configurable block is connected to one stage of the MIPS pipeline. It is reasoned that the re-configurable logic integration over the complete pipeline will not result in more gain compared to the cost of extra logic and routing hardware. The programmable functional units are modelled like multi-cycle instructions. This adjustment solves the latency mismatch issue with fast micro-processors in another method than done in PRISC [60]. The ISA is not modified to retain the compatibility of existing MIPS-compiled binaries.

A more advanced version, termed as OneChip-98 [62], addressed this ISA modelling issue by mimicking PRISC [60]. A particular instruction format is used for custom instructions. A field in the custom instruction points to the memory location storing FPGA configuration bit-stream. Due to the multi-cycle modelling of custom instructions, issues with memory coherence appear here. In OneChip-98, it is noted that, parallel execution of processor and the re-configurable block may be desirable unless there is a dependency between two instructions. These issues are elaborated in

detail along with specific solutions for each. To monitor the instruction
sequence in the FPGA, reservation stations are used.

RaPiD As the scientific interest on binding an FPGA with general purpose
computing kept on increasing, the challenges were mounting on the available FPGA technology. RaPiD [63] attempted to solve this issue. The
challenges were pinpointed to two specific items. Firstly, the available
re-configurable hardware platforms were too generic. Those were good
for random logic implementation but, performed poorly for general arithmetic functions. Secondly, the automated compilation of an algorithm written in high-level language to a fast and efficient implementation was still
missing. RaPiD focussed on the design of coarse-grained re-configurable
block and associated tool-suite. The final goal was to integrate this with a
RISC architecture. The micro-architecture of the re-configurable block is
pipelined. The pipeline stages are segregated by registers providing static
and dynamic control signals. With this organization, the mapping of signal processing algorithms were shown to be efficient [64]. Furthermore, a
suitable language is developed for describing the functions to be mapped
on the RaPiD re-configurable architecture. The language RaPiD-C aided
by the corresponding compiler allows the designer to configure the architecture from an high level of abstraction. In a case study presented at [64],
it is shown that the partially re-configurable processor architectures with
specially designed coarse-grained FPGA blocks are able to strike a good
balance between the ASIC and the DSP implementations.

MATRIX In [65] and [42], the computational density versus functional diversity of processors and FPGAs are compared using definite models. In
[65], it is argued that FPGAs fare excellent for regular data-intensive operations but, perform poorly when high functional diversity is required.
This is due to the scarcity of resources for distributing the instructions locally around the FPGA. To make the FPGAs capable of serving a variety
of functions while retaining its computational advantage over processors,
the MATRIX re-configurable architecture is proposed. The architecture is
fully re-configurable with the Basic Functional Units (BFU) supporting
restricted control logic. The localized control observed in MATRIX was
earlier reported by PADDI [56]. Apart from the control logic, the BFU
consists of a memory and a versatile Arithmetic Logic Unit (ALU). Therefore, the BFU is capable of performing a different variety of tasks in the
re-configurable block. Additionally, the BFU can be pipelined at the input port. Another important design contribution of the MATRIX array is
to have a dynamically switchable network connection. Different network
connections e.g. nearest-neighbor connection, global bus are allowed. Depending on the network routing delay, pipeline delay stage between the
producer and the consumer can be established.

RAW The RAW machine [66] presented a completely new paradigm in processor architecture. The micro-architecture in RAW is proposed to be made
of multiple identical tiles. Each of the tiles contain its own memory file,

register file, ALU and configuration logic. It differed from superscalar processors in the way the memory and register file is shared, which leads to costly hardwired scheduling techniques for the later. It also differed from the existing coarse-grained re-configurable logic arrays by completely doing away with the data bandwidth problem. The re-configurable systems, as proposed by MATRIX [65] or PADDI [56], have local controls. However, the switching interconnect for them is distributed among the local memories. For RAW, the memories are kept separate. However, the interconnects and switching matrix is distributed around the functional units of the tiles. In a nutshell, RAW was a novel proposition attempting to move the issues to the software side and baring everything to the programming environment. RAW micro-processor was hypothesized to be connected with a host processor, managing off-chip memory and streaming I/O devices. The RAW compiler is assigned with a wide range of tasks. The initial compilation phase is to identify parallelism in the code. Following that, placement of the threads to physical tiles is done. Finally, routing and global scheduling of the network resources is done to minimize the program runtime. It is noted in [66] that dynamic support for identification of parallelism or resolving dependencies are necessary. Hardware-based solution of these problems are conceptually different from RAW paradigm. Therefore, several software-based solutions for resolving memory dependence, message passing are outlined. It is also noted that, often hardware implementation of such checks can be more efficient.

Garp   With the architectural organization of partially re-configurable processors heading towards a widely open topic, Garp [50] offered another design point. The ISA of Garp is based on MIPS. The software tool-flow is enhanced to include a configurator, which converts the re-configurable block configurations from a high-level description to a bit-stream. Garp software environment also include a modified MIPS simulator and a modified MIPS assembler. The selection of custom instructions are done manually. A significant research investment for Garp is done to get past the limitations of re-configurable hardware technology. Especially, the effect of run-time re-configuration overhead, memory access from re-configurable hardware and binary compatibility of executables across versions of re-configurable hardware are taken into account. The organization of functions in a re-configurable block is done with clear separation between control blocks (for loading configuration), memory access and data blocks. For repeated loading of recently used configurations, a configuration cache is kept. Specific functions e.g. multi-bit adders, shifters are designed with more hardware than typical contemporary FPGAs. In that regard, Garp made the first step to integrate a coarse-grained re-configurable hardware block in a partially re-configurable processor.

In a further enhancement of Garp, significant research was made to enable plain C code to be executed on a Garp Processor [67]. For a processor-incognizant designer, the C compiler is assigned to perform the following tasks.

1. kernel identification for executing on re-configurable hardware.
2. design of the optimum hardware for the kernels. This includes the module selection, placement and routing for the kernels on to the re-configurable hardware [68].
3. modification of the application to organize the interaction between processor instructions and the re-configurable instructions.

The single-most important contribution of Garp was the development of this compilation tool, named *Gama*. The tool handles several aspects of C to re-configurable block synthesis as outlined above. Gama features novel approaches to reduce the routing delay in coarse-grained FPGA mapping. To take the advantage of run-time re-configurability in Garp processor, the compiler had to solve this mapping, placement and routing problem in a linear-time dynamic programming algorithm. Though the effort was commendable, it admittedly resulted into sub-optimal solution.

NAPA NAPA [69], abbreviated form of National Adaptive Processor Architecture, marked one of the first commercial offerings of partially re-configurable processors. In a major effort to bring adaptive processors to mainstream computing, a system with several NAPA processors were conceived.

The application written for NAPA is automatically partitioned by NAPA C compiler into fixed processor instructions and re-configurable block. Alternatively, the programmer is free to manually migrate portions of C code via NAPA constructs to the re-configurable block. Furthermore, the programmer can resort to an ensemble of CAD tools for programming the re-configurable block and place the appropriate subroutine call in the C program. To improve the complex processor interfacing for high-level programmers, NAPA provides an extensive support for debugging. The fixed processor contained hardware support for breakpoints, instruction tracing and single stepping. The re-configurable block can be also single-stepped all through or for selected portions. The complete processor debugging is accessible to the software debugger via JTAG interface. NAPA also provides a rich simulation environment for tuning the application and the content of the re-configurable logic [70]. The simulator is specifically targeted for NAPA processor. However, it offers early evaluation and benchmarking opportunity for the software developer. The simulator is built by combining a cycle-accurate RISC simulator of the core processor and an event-driven logic simulator for the re-configurable block.

REMARC REMARC [71] (abbreviated form of REconfigurable Multimedia ARray Coprocessor) proposed another processor architecture specifically targeted for multimedia domain. Similar to earlier partially re-configurable processors, REMARC noted that the FPGAs must be coarse-grained and adapted to the target application domain for ensuring better performance. The main focus of the work was to design the re-configurable co-processor.

Like contemporary academic works, the ISA of MIPS [72] is used as the base ISA. Definite instruction extensions for configuration, memory load-store operations are done. For compiling a C code for REMARC, a specific REMARC assembler is developed. The C code can host direct assembly instructions via special directives. For simulation purposes, a multiprocessor MIPS simulator is extended with REMARC functions. The micro-architecture of REMARC re-configurable block is divided into a global control unit and a two-dimensional array of processing elements, referred as Nano Processor. A number of arithmetic, logical and load store operations are supported for each Nano Processor.

Hybrid As pointed out in [73], FPGAs were particularly suited for data-intensive computations but, fared poorly for control-intensive tasks. The purpose of Hybrid was to explore various processor-FPGA combinations targeted for digital signal processing applications. Unlike earlier RISC-based partially re-configurable processors, it is argued in [73] that Digital Signal Processors (DSPs) make a better candidate as the base processor for signal processing applications. The reasons are cited as low-power processing capabilities of DSP, availability of multiple memory banks allowing high data bandwidth and the deterministic nature of DSP applications providing well-defined scheduling of custom instructions for re-configurable block. Especially with high run-time re-configuration overhead and multi-cycle execution of re-configurable block, the scheduling was a challenge in earlier designs. On the other hand, the difficulty of integrating a DSP with a re-configurable block is the unavailability of sophisticated re-targetable tools. At that time, most of the DSPs were manually designed with extensive effort for designing the HLL compiler. The micro-architecture of Hybrid offers a new design point in rASIP design space. To do away with the memory bandwidth bottleneck, the re-configurable block of Hybrid is attached as a co-processor, accessing memory ports independently of the base processor. At the same time, to ensure close coupling, the register file of the base processor is made visible to the re-configurable block.

PipeRench PipeRench [47] approached the problem faced by most of the earlier rASIP designs. The re-configurable hardware blocks available in form of FPGA were slow in re-configuring dynamically. The pipelined re-configurable architecture built by PipeRench supported *pipelined re-configuration* to solve the aforementioned problem. The configuration of parts of the PipeRench architecture can be scheduled along with the execution of other parts. This reduced re-configuration overhead. Alternatively, this is possible by scheduling the re-configuration call in the base processor well ahead of the re-configurable block execution call. The application kernels to be synthesized on PipeRench can be written in a C-like intermediate language, where from automatic place-and-route are supported. The re-configurable hardware elements are chosen to fit the target application kernels with pipelining constraints.

Chimaera A partially re-configurable dynamically scheduled superscalar processor was reported at [74] and [75]. The Chimaera processor design focussed on general programmability by re-targeting a GNU C compiler [76]. The basic ISA in Chimaera is as in MIPS. It is extended with the reconfigurable instructions along with a abstract modelling of their latency. The SimpleScalar [77] tool-set is used to build the simulation environment for Chimaera. The Chimaera compiler is equipped with optimizations like branch collapsing and Single-Instruction- Multiple-Datapath (SIMD). The first optimization enables parallel execution of multiple branches of an application in the re-configurable block. The second optimization leverages the re-configurable block computations of sub-byte range. On top of these optimizations, there is support for identification of application kernels suitable for re-configurable block. The instructions targeted for re-configurable block are associated with a configuration and a unique id. Before the execution of the instruction, the configuration is loaded by initiating a trap instruction. Understandably, this incurs strong penalty performances for repeated loading of configuration bits from memory. This problem, to some extent, is alleviated by having a dedicated configuration cache unit in the Chimaera micro-architecture. To allow out-of-order execution of all the instructions, a scheduler for the re-configurable instructions is maintained. The re-configurable block is allowed to access up to 9 registers in one cycle and write to 1 register after one instruction execution is done. These registers, termed as shadow registers, are partial, physical copy of the host processor's general purpose register file of the host processor. These shadow registers helped to avoid the data bandwidth problem between re-configurable block and the base processor to some extent.

MorphoSys In [78], MorphoSys, a partially re-configurable processor is described. The base processor of MorphoSys is a pipelined RISC one. The multi-context re-configurable array acts as a co-processor. To address the data bandwidth bottleneck, a DMA controller is maintained between the global memory and the re-configurable array. Moreover, the data, which is recently used in the re-configurable array, can be cached in a frame buffer. Several instruction extensions are done in the base RISC processor to enable loading of context, data and broadcast the data and context around the re-configurable array. A new compiler is developed for MorphoSys by re-targeting SUIF [79] frontend. The high-level algorithm, written in C, is to be partitioned manually by the designer using special directives. The code to be executed on the re-configurable array is translated from C to Hardware Description Language (HDL) representation by automatic translators. For verifying the performance of the re-configurable array, an HDL simulator is developed. The micro-architecture of the base processor in MorphoSys is kept simple with major emphasis on the data transfer and instruction decoding for the re-configurable array. The re-configurable array is designed specially for supporting highly parallel and computation-intensive applications. Each cell of the re-configurable array is able to perform several logical and arithmetic operations including multiplication, addition and multiply-accumulate functions. The routing network is designed in three hierarchical levels to provide various levels of data-transfer between cells. Additionally, global buses are designed to broadcast data and context over the complete array. In a recent paper [80], the MorphoSys architecture is evaluated for cryptographic applications, where it advocates the use of local memory elements in the re-configurable array elements.

Chameleon A key motivation for the Chameleon [81] project was to achieve energy-efficiency for mobile multimedia systems. It was realized that with limited battery resources and increasing performance demands, the systems must be less power-hungry. Furthermore, to cope with the changing requirements, it is prudent to have a strong adaptability. Chameleon is essentially a SoC, where coarse-grained re-configurable processors are the building blocks. The key tile processor, named Montium, is targeted towards digital signal processing algorithms. To achieve energy efficiency, the control overhead is minimized and most of the scheduling is done statically. This required a high-level compilation framework. As part of this, a mapping algorithm from Directed Acyclic Graph (DAG) to Montium tile is developed [82]. Important observations are made from the overall Chameleon project in [81]. It is mentioned that to achieve energy-efficiency, locality of reference acted as a guiding principle. Instead of communicating the input and output data throughout the design, it is better to store the relevant data close to the functional unit. The importance of striking the right balance of flexibility and specialization is also pointed out. Several design decisions are cited to show how it affected the performance or over-achieved the flexibility. The lessons clearly indicate the need of an architectural exploration for this complex design space.

Zippy In [83], the design parameters of partially re-configurable processors are listed as following.

1. Integration of host processor and the re-configurable block. This includes the coupling level, ISA and the operands.
2. Re-configurable unit parameters. This consists of the logic granularity, interconnect topology and the re-configuration methodology i.e. configuration caching, static/dynamic re-configuration.
3. Programming model, which is about the tool-flow for software and hardware implementation.

To handle this complex set of parameters, a design methodology for systematically designing and evaluating re-configurable processors is proposed in [83]. Accordingly, an architectural simulation model, termed Zippy, is proposed in [84]. The simulation environment for Zippy is a based on a co-simulation model. The software tool-flow includes a modified GNU C compiler [76], a place and route tool for mapping data flow graphs to Zippy architecture model. The Zippy micro-architecture is organized as per the loosely coupled co-processor model. The parameterizable re-configurable array of Zippy is built as a multi-context architecture. Several configurations for the re-configurable array can be stored in the configuration memory. The re-configurable unit can switch quickly between different configurations. To minimize the re-configuration overhead, a dedicated unit called *context sequencer* is maintained. To increase the computation density over a limited re-configurable area, Zippy proposed a hardware virtualization technique, which was earlier used by PipeRench [47].

XiRISC XiRISC [85] marked one of the early re-configurable processors with a VLIW paradigm. The ISA of XiRISC is RISC-based with special extensions for the re-configurable array. Two types of processor instructions are appended to the basic ISA namely, instructions for loading the configuration and instructions for executing it. To explore the ISA extensions, while quickly re-targeting the complete simulator, XiRISC resorted to ADL-based tool-suite generation [86]. The base processor ISA is completely modelled using ADL LISA [10], while the re-configurable block is emulated via dynamically linked libraries. For debugging purposes, the verbose version of the emulation library can be used. XiRISC simulator [87], for the first time, offered a comprehensive and re-targetable simulation environment capturing all the effects of the re-configurable hardware as well as the ISA. Due to the emulator environment, no high-level modelling of the re-configurable operations' latency [74] is required anymore. The emulator library is designed to be completely target-specific. No methodology to explore the re-configurable array is specified.

The runtime re-configurable array in XiRISC is organized in the micro-architecture like any other VLIW slot with a capability to access general purpose registers, memory and other internal processor resources. The re-configurable array is designed to perform operations with multi-cycle latency. To increase the throughput, the complete re-configurable array is pipelined. The control sequence of the pipelined re-configurable block is extracted from the C-based application kernel by instruction scheduling. The re-configurable logic cells are fine-grained in nature with special extensions for fast carry propagation. To reduce the re-configuration overhead, configuration caching as well as multi-context decoding is supported.

MOLEN The major challenges of rASIP design, as identified in [51], were firstly, the instruction space explosion and secondly, the limited bandwidth between base processor and the re-configurable block. MOLEN proposed a polymorphic instruction-set paradigm, which is essentially a one-time instruction-set extension accommodating a wide number of possibilities of re-configurable instruction. The extended instruction-set included instructions for loading and pre-fetching the configuration, executing and pre-fetching the execution microcode, load-store instructions and a `break` instruction. The break instruction can be used to synchronize the host pipeline and the re-configurable block executions. The load-store instructions (`movtx`, `movfx`) are used to move data between the host pipeline's general purpose register file and the special register file dedicated for the re-configurable block. By these special registers, re-configurable block is supported with a high data bandwidth. However, a number of instructions are spent for data movement, thereby possibly achieving the same performance improvement as previous rASIP designs. The pre-fetch instructions for configuration and execution are used to prepare the re-configurable block for execution well ahead of the actual re-configurable instruction triggers. To take the complete advantage of the MOLEN extended ISA and show admirable performance gain, an excellent HLL compiler is required. The SUIF [79] compiler front-end is re-targeted for PowerPC processor [88], which acts as the base processor for prototype MOLEN implementations. An instruction scheduling algorithm for MOLEN is implemented in the compiler [89]. Another interesting work in this context is to allocate the FPGA area among target application kernels, so as to minimize the dynamic re-configuration overhead while maximizing overall speed-up performance [90].

Others Several partially re-configurable processors are designed and presented at [91, 92]. The major contribution of these designs is to come up with a FPGA template tailored for arithmetic operations. The FPGA template can be parameterized for input-output ports of various logic elements, alignment of logic elements in the cluster, number of routing channels and segmentation for the interconnects etc. These parameters can be passed to a *Datapath Generator* for obtaining a fully placed and routed custom FPGA. A simple mapping tool is designed to map applications to an arbitrary FPGA macro. The mapper-generated bitstream and the FPGA layout is passed to existing commercial tools for obtaining realistic simulation results on power, area and timing. In [91], the custom FPGA is demonstrated to outperform commercial FPGAs in arithmetic-oriented applications. In [92], the custom FPGA is attached with two different RISC processors. In a noteworthy effort, an existing VLIW-based partially re-configurable processor is compared with these proposed designs for cryptographic and multimedia applications.

3.2 rASIP Design : High-level Modelling Approach

The high-level rASIP modelling, albeit in a limited form, is attempted by the following methodologies.

Pleiades In [46], it is pointed out that re-configurable computing offering faster computation came at the cost of increased area or power. This necessitated a design space exploration with accurate cost models. With the era of embedded applications heralding, Pleiades put the focus on designing low-energy re-configurable processors for digital signal processing and multimedia applications. Pleiades provided the first framework of architecture exploration for partially re-configurable processors.

The architecture template of Pleiades consists of a programmable base processor and heterogeneous computing elements, referred as *satellites*. The communication primitives between the architecture and the satellites are fixed by this template. For each algorithm, an architecture instance from this template can be created by specifying the number of instances for each satellite. Essentially the satellite works as an accelerator. The satellite instances can be Application-Specific Integrated Circuit (ASIC) and/or FPGA or a mix of both. The satellite and the micro-processors are initially characterized with power-delay-area models. The algorithm is mapped to a microprocessor and then computation-intensive kernels are extracted from that. Standard gcc-based profiling tools are extended to identify the computation kernels. The power-delay-area cost of mapping a kernel to a satellite are computed analytically on the basis of the initial characterization. Depending on the evaluation results, semi-automatic partitioning of the application are performed. For each mapping of a kernel to a satellite, a kernel-specific simulator can be instantiated. A further exploration of the re-configurable block interconnect architectures can be performed to minimize power. It is noted in Pleiades that, the interconnect architectures consume most of the power in a re-configurable hardware. The most prominent interconnect architectures namely, multi-bus, irregular mesh and hierarchical mesh structures can be evaluated for a given application using Pleiades framework.

ADRES The most recent work on a high-level rASIP exploration is reported in [34, 93, 94]. The ADRES architecture is essentially a combination of a VLIW base processor with a re-configurable array of functional blocks. Contrary to previous RISC-based [35, 71, 46] and DSP-based [73] rASIPs, ADRES argued that the VLIW host machine allows the designer exploit the limited parallelism available in the application, which cannot be mapped to a re-configurable array. Furthermore, the VLIW alignment of the functional blocks releases the pressure of data bandwidth between host processor and the re-configurable block faced by almost all earlier rASIPs. Conceptually, the task of loading data to the parallel re-configurable functional blocks are moved upwards to the software programmer. This entailed a robust compiler framework for ADRES architecture instances. The software programming environment of ADRES is essentially based on an existing VLIW compiler framework [95]. A single ADRES architecture instance can be simulated in a co-simulation environment. During architecture exploration, the designer is allowed to investigate the effect of following parameters on the performance.

1. Number of functional units. One functional unit is instantiated in one slot of the host VLIW processor. Each VLIW slot again connects to a row of functional units, resident in the re-configurable block.
2. The size of register file. One register file can be associated with every functional unit in the re-configurable block. This leads to energy-efficiency as observed earlier in Chameleon [81].

3. The number of register files.
4. The operation set supported by the functional units.
5. Latency of the target architecture.
6. Interconnect topology of the re-configurable array.

The aforementioned parameters can be specified in an architecture description in XML format. The architecture description is used to generate the simulator as well as the compiler [96]. During compiler generation for a particular instance, an earlier VLIW parallelization technique [97] is adopted. The novel modulo scheduling algorithm employed in [96] considers not only the available parallelization but also, the limits imposed by the interconnect topology and the resource constraints in the re-configurable array. To demonstrate its modelling capability, MorphoSys, an earlier rASIP [78], is modelled using ADRES.

3.3 rASIP Design : Library-based Approach

KressArray KressArray [98] offered the first design space exploration environment for fully re-configurable architectures with the ability to select from a library of architectures. The KressArray design space include the following items.

- The size of the re-configurable array.
- The functions covered by the re-configurable Datapath Units (termed as rDPU).
- Number of nearest-neighbor ports.
- Interconnect topology of the re-configurable array.
- The number of row and column buses.
- Areas with particular rDPU functionality i.e. specific area of the re-configurable array can be designed to contain a certain rDPU function, different from other areas.
- The number of routing channels through a rDPU.
- Location of peripheral ports and their grouping.

For design space exploration, the designer needs to write the application using a high-level ALE-X language. From this language, internal tools estimate the minimal architecture requirement and trigger the designer-driven exploration process. With each design cycle, the designer can update the architectural parameters and perform mapping and scheduling [99]. The mapping of the application to the architecture instance is driven by a simulated annealing-based algorithm. The scheduling phase decides precise start time of each operation in order to satisfy the resource constraints for providing data from the main memory. Several high-level optimization steps are performed before mapping to increase parallelism in the application kernel. The output of these tools are in the form of statistical data, providing valuable feedback. The application and the architecture can be simulated by a simulation environment, too. Once an optimum re-configurable array is defined, the Verilog HDL representation of that can be generated automatically.

Stretch Stretch [33], a natural extension of library-based processor modelling approach of Tensilica [29], marked the first commercial offering of library-based partially re-configurable processor modelling environment. Stretch's S5000 family of software programmable processors consist of the 32-bit Xtensa-based processors offered by Tensilica [29]. With that, Stretch embeds a coarse-grained re-configurable block, termed as Instruction-Set Extension Fabric (ISEF). To alleviate the data bandwidth issues, the ISEF is connected to the base processor and memory via 128-bit wide register ports. The register file dedicated for the ISEF is itself organized as 32 128-bit wide registers, whereas the base processor is aided with a separate 32-bit wide register file. For S5000 family of processors, 3 input operands and 2 output operands are allowed for the ISEF. Stretch provides an advanced GUI-based software programming environment referred as Integrated Development Environment (IDE). The IDE consists of a cycle-accurate simulator for early performance tuning, Stretch development board for system verification and an in-circuit debugging environment via JTAG interface to the target system. The identification and extraction of application kernels are supported by Stretch C compiler. The compiler is able to automatically identify program hot-spots, generate the configuration code for the ISEF and replace the hot-spot via the custom instruction calls. Recently, another family of partially re-configurable processor family (S6000) is announced by Stretch. In S6000, the ISEF is equipped with an embedded RAM for storing heavily used data locally. The processor is also extended with several dedicated programmable accelerators.

The huge amount of design space, as extended and explored by aforementioned designs and tools, makes the task of establishing a generic design methodology extremely challenging. To head of a wholesome approach of designing rASIPs, a collection of design points is required. Several surveys and classifications for re-configurable processors are done in [48, 49, 100, 101]. In the following Fig. 3.1, a representation of the complete design space is drawn without broadly classifying the processors in any category. The design space is essentially a hyperspace with multiple intertwined design dimensions. The dimensions are represented with the arrows. The design points are mapped on the arrows following an outward evolutionary path i.e. the innermost points represent earliest designs. As can be easily understood, the current rASIP design approaches cover the complete design space partially.

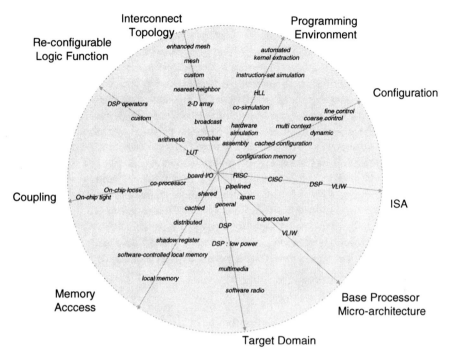

Fig. 3.1 Partially re-configurable processor design space

3.4 Independent rASIP Design Tools

The overall rASIP design flow necessitates several tools, which can be used in other contexts than rASIP. The most prominent of these is a high-level profiling environment for characterizing the application domain. A follow-up tool of high-level profiling is the kernel extraction and mapping tool, which identifies the hot-spots in the application and maps that to the processor or the re-configurable block. Apart from several tools integrated into the aforementioned rASIP design environments, an impressive research work is performed in those two areas independently. The following section outlines major steps in those areas.

3.4.1 Profiling

Several standard tools exist for source code level profiling. These tools (e.g. GNU gprof [76]) are used by software developers for obtaining statistics about the CPU time consumed by different functions. The profiling results can be used to modify the C/C++ code with the target processor in mind. The profiling results obtained in high-level abstraction can be misleading when a rASIP / ASIP is considered. This led to the development of target-specific profiling techniques in existing processor designs or processor design frameworks. There the profiling is performed at the machine-specific assembly code level. Such profilers are mostly machine-specific (e.g. SpixTool [102] for SPARC or VTune [103] for Intel processors). In ADL-based processor design approach, re-targetable assembly-level profilers are embedded. This kind of profiling is referred as fine-grained profiling compared to the earlier high-level coarse-grained profiling at C/C++-level. Fine-grained profiling requires an initial architectural implementation and delivers a high accuracy of results. Whereas, coarse-grained profiling can be done in a machine-agnostic manner. But, the profiling accuracy is compromised. Several profilers are presented to strike a balance between profiling accuracy and target-independence. The SIT toolkit [104] presents an approach to obtain fine-grained profiling results by using operator overloading techniques. A novel code instrumentation technology and consequently profiling at Intermediate Representation (IR) level is reported in [105]. Due to the profiling at IR, the high-level C optimizations can be enabled. This avoids several potentially misleading statements, encountered in C-level coarse-grained profiling approach. In another extension of this work, a capability to profile dynamic memory access is shown [106].

3.4.2 Automated Kernel Synthesis

Recent interest in the design of programmable processors have increased the motivation of automatic kernel extraction. In most general form, this kernel is mapped to an ASIP instruction or synthesized on the re-configurable hardware. In either case, the task is to determine the part of an application, which will fit the target architecture without costing much overhead in synchronization. A significant amount of research in this area is performed during the evolution of partially re-configurable processors [99, 35, 34, 33]. In the following, the independent research done in this field are briefly reported.

Typically, the kernel extraction is performed in a high-level language framework. The problem is attacked by constructing a DFG for the basic blocks of application, where the nodes represent the operations and the edges represent the data dependency. Clearly, the solution quality of this problem is largely dependent on this DFG construction. By performing various compiler optimization, the size of the DFG can be increased. A larger DFG, in turn, allows more scopes of finding suitable kernels. The kernel extraction from the DFG focuses on two things. Firstly, the sequence of atomic operations, which can be mapped onto pipelined re-configurable hardware. Secondly, the operand constraints. More input/output operands allow a larger DFG to be mapped onto re-configurable hardware at the cost of increased data bandwidth, which must be supported. A greedy approach presented at [107] can determine the maximal Multiple Input Single Output (MISO) patterns. An iterative algorithm exploiting instruction level parallelism for VLIW architectures is reported at [108]. This method can detect Multiple Input Multiple Output (MIMO) patterns. A significant extension over [107] is described in [109]. The paper describes a branch-and-bound algorithm which selects maximal-speedup convex subgraphs of the application dataflow graph under micro-architectural constraints. This approach can identify MIMO patterns. An improvement over [109] is presented at [110]. This algorithm partitions the DFG into several regions and finds the patterns within each region. By this approach, strong runtime improvement over [109] and better scalability is achieved. More recently, a heuristic algorithm with polynomial-time complexity is reported by [111]. In this heuristic, a MIMO DFG pattern is identified within a basic block. The computational complexity of this heuristic is reported to be $\mathcal{O}(N^{I+O})$, where I and O are the cardinality of the input and output sets respectively. In an interesting development of these approaches, an Integer Linear Programming (ILP) formulation is presented for the first time by [112]. Using this ILP formulation, it is shown that, the runtime is much better than the previously presented heuristics. Another ILP-driven algorithm is presented in [113]. The importance of this approach is that, it handles several designer-specified area, latency as well as architectural constraints. Being targeted towards a flexible ADL-based [17] back-end, this algorithm can be potentially used for any processor with a fast design space exploration loop.

With ILP-driven approaches offering low-runtime identification of large DFG patterns, the other important research problem remains is to integrate the identified custom instructions to the base processor without much cost overhead. It is reported at [114] that by setting extremely high input-output bounds it is possible to identify custom instructions with high runtime speed-up potential. However, high number of inputs and outputs may not be supported due to interface restrictions. Therefore, in [114] the data are moved to the local register file of the accelerator block hosting custom instructions. The presence of local memory elements in the re-configurable block is exploited in [115]. There, the genetic algorithm-based approach tries to identify larger DFG and places the data movement to the local memory before the kernel. Another algorithm to release the data bandwidth pressure by employing shadow registers while identifying custom instructions is presented in [116]. In [117], the register access is distributed over the complete pipeline to reduce the number of register ports for high input-output DFG pattern. Another work reported in [118] tries to utilize the pipeline data forwarding paths as inputs to the custom instructions.

In an important cost-driven approach of custom instruction identification, an automatic methodology is presented at [119]. For each of the potential custom instructions, hardware mapping on the target processor is done to estimate the impact. The impact of high-level optimizations on instruction-set customization is reported in [120], where arithmetic optimizations and symbolic algebra are used for enhancing the performance improvement potential of a kernel. Applications with a high number of arithmetic statements benefit significantly from this approach. Another processor-dependent approach for kernel synthesis is presented at [121]. The significance of this approach is that, the complete kernel extraction and synthesis is performed dynamically. During compilation, hints about the potential kernels are embedded into the code.

An ASIP design methodology for the control-dominated domain is proposed in [122]. This paper uses the ILP technique for synthesizing kernels to custom instructions. The overall task is divided there into two decoupled optimization problems. Firstly, the operator sequences are identified to meet the timing constraints. Secondly, the number of parallel instruction issues are reduced by grouping pairs of instructions. For embedded systems with real-time constraints, the instruction set extensions must be identified with worst case execution scenario. This is done in [123].

For rASIPs, it is also important to re-use the custom instructions as much as possible. In [114], the template custom instructions are first identified and those are matched over the complete application for finding opportunity to re-use that. The template matching is performed via checking graph isomorphism.

In the context of *dynamic re-configuration*, it is important to re-visit the kernel synthesis algorithms, as investigated in [124]. It is shown there that, for a fine-grained FPGA (Xilinx Virtex 2 Pro), the re-configuration time for the entire FPGA can be quite high (ranging from 20 ms for the device XC2VP20 to 47.55 ms for the device XC2VP50). To recuperate from the losses in the dynamic re-configuration, various strategies for the kernel synthesis can be adopted. Obviously, the simplest is to statically re-configure it. However, the application may require multiple custom instructions, which cannot be fit together in the pre-execution mode. In such a case, the FPGA can be partitioned and separate area for each custom instruction can be allocated. The temporally exclusive custom instructions can be mapped on the same area as done in [124]. This requires, however, the FPGA to have partial re-configuration capability. The dynamic re-configuration overhead is comparatively less for coarse-grained re-configurable fabrics. Even then, when connected to a base processor, this accounts for few thousand cycles [125] for a processor running at hundreds of MHz. In [125], a dynamic programming-based algorithm is proposed to obtain the optimal spatial (for spatially exclusive custom instructions) and temporal (for temporally exclusive custom instructions) partitioning.

3.5 Motivation

The concept of flexibility was introduced with the expansion of design space of digital systems. The most flexible digital system consists of a GPP realized on a post-fabrication customizable hardware. When these systems are targeted for selected application domains, they perform poorly compared to say, an ASIC targeted exclusively for that application domain. On the other hand, exclusively designed ASIC incur high NRE cost and long design cycles due to poor flexibility. The art of system design is essentially to perform trade-off between the extremes of specialization and generality. Application-Specific Instruction-set Processors (ASIPs) offer a fine balance between generality and specialization. With ASIPs, flexibility for an application is retained by having it software-programmable and specialization is achieved by special micro-architectural features. With the ASIPs being partially re-configurable, another knob to engineer the trade-off between generality and specialization is established. More importantly, the flexibility of the ISA and the computational organization can be retained even after the fabrication.

A major aspect of the modern system design is to achieve longer time-in-market in the view of increasing NRE cost. With the fast evolution of application areas such as signal processing and multimedia, it can be conjectured that software programmability alone may not be sufficient for a system to deliver high performance over long time. Controlled post-silicon flexibility, in form of re-configurable hardware, appears to be a judicious approach. From a technical point of view, an ASIP with post-silicon flexibility allows maximum opportunity to extract the performance over the complete evolution of target application(s).

The lack of a convenient modelling framework restricted most of the early rASIP designers to remain limited within a few design decisions. The early rASIP designs plainly demonstrate the evolution of rASIP design space with each design adding new design points. Only the recent rASIP designs approached the overall design space by performing some design space exploration. This is done either via library-based modelling approach [33, 93] or via abstract modelling approach for a part of the complete tool-flow [87]. The only methodologies to offer abstract model-based rASIP exploration are Pleiades [46] and ADRES [93]. The Pleiades methodology concentrates on a heterogeneous system-on-chip model with specific micro-processors as base and a definite architecture template. Without the initial set of well-characterized micro-processors and re-configurable blocks, the design space exploration can not be initiated. Furthermore, no design space exploration methodology for overall ISA and/or the FPGA basic blocks are specified. The ADRES architecture template is modelled as per VLIW paradigm. The design space exploration is limited by the template constraints.

In this book, the abstract processor modelling approach is extended to cover the complete rASIP design space. The ISA modelling, software tool-suite generation, re-configurable hardware modelling and the automatic rASIP implementation are done in a fully generic manner. This enhanced degree of freedom enables the processor designer to do the best performance-flexibility trade-off. However, for fast rASIP implementation or re-using legacy IP, the design space can be suitably curbed. In other words, the proposed language-driven rASIP exploration framework provides a common platform for every processor designer without any loss of generality.

3.6 Synopsis

- rASIP designs evolved with early board-level FPGA-processor coupling to latest custom FPGAs residing within processor as a functional unit.
- The designs attempted to blend the contrasting advantages of processor and FPGA, while addressing the discrepancies via software or hardware mechanisms.
- Regardless of design choices, a significant gain for the rASIPs in terms of performance over classical processors are shown.
- Due to the huge and complex design space of rASIPs, the design space exploration and trade-off are often limited within a sub-space.
- A high-level design space exploration and implementation framework based on abstract processor modelling approach is proposed in this book.

Chapter 4
rASIP Design Space

The whole is more than the sum of the parts.
Aristotle, Philosopher, 384–322 BC

In this chapter, the design space of rASIP is elaborated from an architectural perspective. For each design point, the effect of other design points are highlighted. Consequently, the design space is mapped to the modelling environment. The complete modelling environment, which includes an extended ADL, is described formally and using suitable examples. On the basis of the proposed rASIP description, the overall rASIP design methodology is presented.

4.1 Architecture Design Points

The classification of processor architectures has been suggested in various literatures [126, 127]. A broad separation, from an architectural perspective, is suggested in [126]. This classification, widely referred as Flynn's taxonomy, is as following.

- *Single Instruction, Single Data stream (SISD)* : This refers to a scalar processor without any spatial parallelism.
- *Single Instruction, Multiple Data streams (SIMD)* : This refers to a vector processor. These kind of processors have the ability to compute on multiple data streams for a single instruction stream.
- *Multiple Instruction, Single Data stream (MISD)* : This is a rather uncommon combination as, for any practical multiple instruction processor architecture, multiple data stream needs to be present. Theoretically, a vector processor without any parallel data processing belongs to this class.
- *Multiple Instruction, Multiple Data streams (MIMD)* : This refers to multi-processor systems.

In this work, we concentrate on uni-processor systems and therefore, we restrict our design space discussions within SISD and SIMD architectures. Obviously, the results can be extrapolated to the realm of MISD or MIMD architecture classes.

For a finer understanding of the interaction between various architecture choices, we divide the rASIP design space across ASIP architecture, re-configurable block architecture and the interface. This is shown in the following Fig. 4.1. In the following sections, each of these architectural sub-spaces are discussed.

A. Chattopadhyay et al., *Language-driven Exploration and Implementation of Partially Re-configurable ASIPs*, DOI 10.1007/978-1-4020-9297-8_4,
© Springer Science+Business Media B.V. 2009

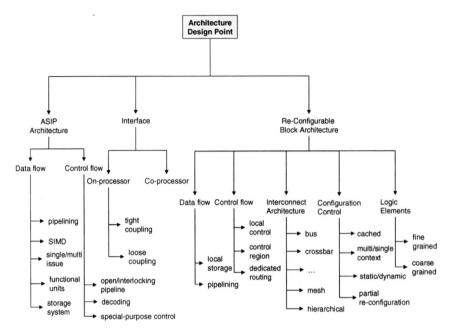

Fig. 4.1 rASIP design points

4.1.1 ASIP Architecture

Within the ASIP architecture two overlapping design spaces exist. Those are the data flow organization and the control flow organization. Within each of these, numerous design points are present. In the following, the salient design points and their influence on the overall rASIP design decisions are discussed.

4.1.1.1 Data flow

Pipelining : To benefit from the temporal parallelism available in the application, pipelining is almost standardly used in modern micro-processors. In pipelining, each instruction is divided and mapped to different parts of the micro-architecture. The instruction passes through these parts sequentially. By having separate parts of the micro-architecture (referred as stages of the pipeline) performing separate tasks, multiple instructions can execute in parallel. Once the pipeline is loaded fully with the instructions, the throughput increases although the instruction latency remains intact. The following Fig. 4.2 shows an example of pipelined data flow. In the figure, the diagonal dotted arrow indicates the flow of an instruction over the complete pipeline.

The pipelining of the data flow includes several design decisions of its own.

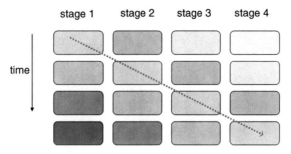

Fig. 4.2 Pipelined data flow

Firstly, the distribution of hardware tasks of an instruction into pipeline stages. This distribution needs to be generic enough for allocating some task for all instructions in each pipeline stage. Otherwise, instructions pass through stages without performing anything. This decreases the area efficiency of the entire design. The distribution of hardware can be too dense for being accommodated in one single stage resulting into long critical path. In such cases, the pipeline is elongated by dividing the corresponding stage.

Secondly, several sequential instructions running into the pipeline may have a data dependency between them causing a hazard in the data flow. The instructions, typically, read the data in a particular stage and write the data after computation in another specific stage. The stages, which lie in between these two stages may contain obsolete data, loaded intermediately before the writing is performed. To solve this issue, the dependent instructions can be scheduled in a manner to avoid this. Alternatively, more flexibility in reading and committing of data in the pipeline can be introduced. More about these techniques can be found in standard literature [128].

Thirdly, the pipelining decision may need to take into account the reading and writing of data from/to synchronous memories as well as cache miss penalties. This necessitates the address generation, request transmission and the data access. Without proper distribution of these tasks across the pipeline stages, several memory-independent instructions might become stalled in the pipeline.

Effect on rASIP Design : The re-configurable block can be conceived as a single functional unit within one stage of the base processor pipeline. Additionally, it should be noted that the same operation may result in higher latency when mapped to FPGA compared to ASIC. Under such circumstances, the re-configurable block dictates the critical path of the overall architecture, slowing down the clock speed and reducing the performance improvement (if any). Alternatively, the re-configurable block can be considered as a multi-cycled functional unit within the span of one pipeline stage of the base processor. In this case, either the pipeline needs to be stalled (shown in the following Fig. 4.3) to ensure the completion of the re-configurable block execution or the hardware mechanism must be employed to allow independent instructions to go ahead. There can be another kind of organization,

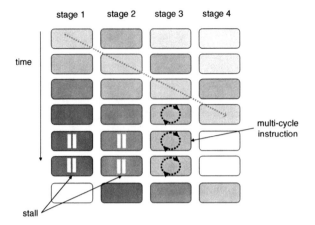

Fig. 4.3 Multi-cycle re-configurable instruction

where the re-configurable block itself is pipelined. This case is discussed in the data flow organization of re-configurable block in detail. As far as the base processor pipelining is concerned, the re-configurable block needs to be synchronized with it. The synchronization options include single-stage single-cycled execution, single-stage multi-cycled execution or multi-stage multi-cycled execution. The options offer different design points in the area-time trade-off.

SIMD VLIW and Non-VLIW: Very Long Instruction Word (VLIW) organization can be applied orthogonally to the pipelined data flow. VLIW is a technique to exploit the spatial parallelism present in the application. The parallel functional units of VLIW are known as *slots*. Each slot receives a specific instruction to decode and execute. To organize the data coherence between the slots, the storage system is distributed across the slots. Again for each slot, a specific portion of storage can be reserved, leading to clustered storage concept. Some of the functional units are replicated and present in each slot, whereas some functional units are present in selected slots. This organization calls for a compile-time analysis of the instructions, thereby exploiting as much Instruction-Level Parallelism (ILP) as possible in view of the given VLIW organization. The following Fig. 4.4 provides an abstract view of VLIW data flow organization. A general SIMD implementation is also possible without having long (and different) instruction as in VLIW. In this case, the instructions have the decoding logic centralized across the slots, only issuing the control signals to each functional unit. According to this control logic, the data is processed in parallel by several homogeneous or heterogeneous functional units. For such an organization it is necessary to despatch data in parallel to the functional units, requiring the data to be aligned.

Similar to the pipelined data flow organization, the partitioning of processing elements into slots calls for an understanding of the target application domain. The division of slots needs to be done to allow the applications to run with high parallelism and at the same time, without wasting slots during execution. A free slot in

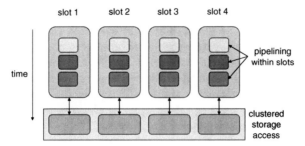

Fig. 4.4 VLIW data flow organization

a large instruction word is usually filled up with *no operation* (nop). This results into empty data propagation across registers and useless decoding, decreasing area-efficiency i.e. under-utilizing the processing resource. The scheduling and packing of parallel instructions in VLIW is performed statically by the compiler.

Effect on rASIP Design : Apart from any other issues, the VLIW data flow organization addresses the single biggest challenge of rASIP design i.e. the data bandwidth between the processor and the re-configurable block. The re-configurable block in a VLIW data flow organization can be arranged vertically or horizontally to the slot organization. In one case, the re-configurable block acts like a slot (or part of a slot). In another case, the re-configurable block is composed of several slots to match with the base processor organization. The two different organizations are presented in the Fig. 4.5(a) and (b). In both the cases, the storage organization of VLIW need to be done in order to synchronize with the re-configurable block. On another note, the task of detecting parallelism and bundling the instructions into a long instruction word needs to be done statically. This issue, existing in any VLIW processor, becomes even more challenging in case of rASIPs, particularly due to the different nature of data path organization. For the re-configurable block acting as a slot, the synchronization of data flow is difficult in presence of the globally

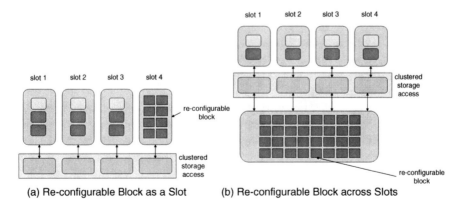

(a) Re-configurable Block as a Slot   (b) Re-configurable Block across Slots

Fig. 4.5 VLIW data flow organization with re-configurable block

accessible storage. This is due to the inherently longer (possibly unknown during compilation time) latency of the operations mapped to the re-configurable block. For re-configurable blocks laid across the slots, this problem is faced differently. For this organization, the processing elements in the re-configurable block need to be fine-grained enough for remaining flexible over the range of instructions processed in every slot. This might require a high number of pipeline stages in each slot to process a complex instruction. For the processing elements being coarse-grained, the rASIP runs the risk of being too target-specific.

For the re-configurable block arranged to process the complete data flow (i.e. in perpendicular to the base processor flow) this offers an interesting data point, as demonstrated in [34]. For the re-configurable acting as a slot, an implementation is proposed in [85].

Single/Multi Issue : Multiple issue slots in the processor hardware presents another way of SIMD organization. While the SIMD organizations presented earlier rely on the HLL compiler for scheduling and packing of the instructions, the multi-issue superscalar processors perform this dynamically via hardware. This extra hardware contributes significantly to the silicon area and energy overhead. Consequently for embedded computing, superscalar processors are rarely used. As our primary focus is in embedded computing domain, an elaboration of superscalar processor design space is omitted here.

Functional Units : The data path of a processor is modelled using a number of functional units. The functional units can have various design choices as listed below.

- *Operation Granularity :* The operations can have various granularity ranging from bit-level to word-level or even higher.
- *Parallelism :* The functional unit can offer parallelism within it. A further design choice is the number of parallel data flows allowed by it.
- *Functionality :* This is dictated by the target set of applications. The signal processing algorithms often require subsequent multiplication and accumulation for filtering. In this case, it is useful to have this operation (multiply-accumulate or MAC) in a functional unit.

Effect on rASIP Design : It is important for a rASIP designer to partition the functionality between the base processor and the re-configurable block prudently. The functional units present in the base processor should not be susceptible to future changes in the application.

Storage System : The state of the processor is defined by the storage system. In the following Fig. 4.6, a generic storage element hierarchy is presented. As shown, the smallest and fastest memory is organized in form of a register file within the CPU. The register file is typical of modern micro-processors, which follow RISC-like load-store ISA. In CISC architectures, direct memory-based computations are observed. The storage elements following the register are arranged in the order of increasing size and access time. After the register, the cache memory is located followed by the main memory.

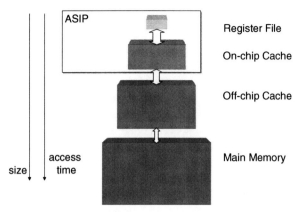

Fig. 4.6 Generic memory hierarchy in ASIP

The storage system design choices with the factors influencing the choices are presented below.

- *Size, width and organization of the Registers :* The target applications are studied to identify the working set and corresponding register requirements. This can be used directly to define the register size. Generally the register file size is fixed to an integral power of 2 to utilize the full scope of address bits. The number of registers is also driven by the requirement of available ports and the size of each register. For aiding temporal parallelism as in pipelined architecture, intermediate special-purpose registers for carrying pipelined data are used. For spatially parallel functional units as in VLIW architectures, multiple clustered registers or banked registers are used. Apart from these, several special-purpose registers for custom functional units, ALUs are reserved.
- *Register Access Pattern :* The special-purpose dedicated registers have straightforward access pattern from the corresponding functional units. The general purpose registers are designed to have minimum number of ports without compromising performance. Higher number of ports result in increased area overhead due to multiplexing but, offers increased parallel operations on data. Especially for SIMD architectures this is important. There, it is achieved by partitioning the register into several clusters. Each parallel data flow is allowed to access to one cluster of the entire register file. Another technique is to divide the register file in several windows (contexts), where the currently executing functional unit can only access one window [129].

Another important classifier for the access pattern is the commit point of data in the processor. Especially for pipelined data flow, the temporal separation between access point and committing point directly controls the cause of data hazards. This can be avoided by bypassing the data between intermediate pipeline stages.

- *Size, width, organization and location of the Memories :* The overall memory subsystem can be unified for data and program (von Neumann architecture), can be completely separated (Harvard architecture) or partially unified (modified harvard architecture). These design decisions largely rely on required program and data space and access patterns. The memories can be located on-chip or off-chip depending on required access speed, shared or distributed nature of memory and area-energy considerations. The cache memory is updated dynamically during the program execution to accommodate the latest data or instruction. For computation-intensive applications, an alternative organization is achievable by user-defined scratchpad memory [130].
- *Memory Access Pattern :* The memory (data and/or program) can be single or dual ported. Each port can allow either read access or write access or both of those. Furthermore, for increasing parallelism the memories can be banked.

Effect on rASIP Design : Data communication being a prime focus of rASIP design, the memory subsystem needs to be designed with extreme care. To avoid the re-configurable block from being starved of data, enough communication ports must be provided for data transmission between the memory subsystem and the re-configurable block. The choices can be as following.

Firstly, the base processor may act as a communication media between re-configurable block and the memory. This preserves the load-store nature of the RISC ISA without increasing the memory port requirements. In this case, the data is loaded from memory to registers, which in turn is passed on to the re-configurable block. This approach usually results in low data bandwidth for re-configurable block. A possible solution to this is offered by VLIW approach, where the register file size is huge [34] or by maintaining special purpose registers for supplying data to the re-configurable block [74, 116].

Secondly, the re-configurable block can have direct access to the cache or main memory. In that case, the memory ports need to be shared between the base processor and the re-configurable block.

Finally, the re-configurable block can contain a local scratchpad memory as well as local register file. For all cases of sharing the memory subsystem between the base processor and the re-configurable block, data coherence needs to be maintained. This problem is especially challenging in case of simultaneous data flow of base processor and the re-configurable block. These possibilities of design space are shown in the following Fig. 4.7. Note that, these possibilities are orthogonal to the data flow organization of the base processor, though several weak and strong influences between those exist.

4.1.1.2 Control flow

The control flow of a processor can be viewed as the basic agent for changing the processor state (preserved in the storage elements) via instructions. The control flow can be sub-divided into the instruction encoding-decoding part and the hazard control part.

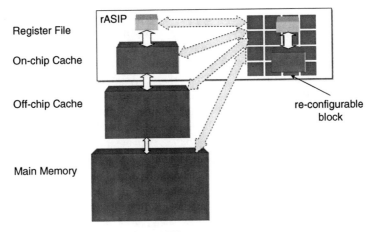

Register File

On-chip Cache

Off-chip Cache

re-configurable
block

Main Memory

Fig. 4.7 Memory access possibilities in rASIP

Decoding : The aspect of flexibility in programmable processors entails the decoding of instruction word. The decoded control signals trigger the functional units to perform necessary computation. For a pipelined data flow, the decoding mechanism can be categorized to either time-stationary or data-stationary one.

In a *time-stationary control* mechanism the control signals for the entire pipeline in a single time unit are generated from a global controller. The complete decoding is performed in one unit and the decoded signals are passed to the execution units in all the stages. In this mechanism, the entire state of the pipeline needs to be handled at a single instance by the controller. The maintenance of all the states consume extra storage area. Nevertheless, this is beneficial for obtaining better layout due to singularly devoted area for irregular control circuitry. Time-stationary control is also easier to design and verify, as those start from a state-machine abstraction of the target application. Hence, it is often preferred among ASIP designers [131].

In a *data-stationary control* scheme, the control of each pipeline stage is propagated along the pipeline and the decoding is done locally. This is comparatively more difficult to design due to the distributed nature of irregular controlling circuitry.

Orthogonal to the above control mechanisms, the instruction op-code plays an important part in the control flow of the processor. The instruction set architecture can be coded with a highly compact code leading to a small instruction-width or can be completely micro-coded resulting in a very wide instruction.

Effect on rASIP Design : Due to the irregular nature of control signal routing across functional units, local control flow within re-configurable block is limitedly applied. This requires the derivation of control signals for re-configurable block within the base processor itself, which can have data-stationary or time-stationary control. If the re-configurable hardware is considered to be a part of the pipeline stage, even then the data-stationary control part of this stage can reside in the base processor. This also indicates that the interface between the base processor and the re-configurable block is fixed with the particular control signals, thereby limiting the flexibility of the design. The alternative to this is to pass the instruction word within the re-configurable block. This allows data-stationary control inside the re-configurable block at the cost of irregular routing paths.

As per the instruction encoding is concerned, it is important for the rASIP to have instruction encoding reserved for the future ISA extensions. This can be done by specifying a group of op-codes for the future instructions. This group of op-codes also determine the upper limit of the number of instructions, which can be added to the re-configurable block. This op-codes can have definite placeholders for instruction operands or can remain open to the arrangement of those.

Interlocking : A pipelined data flow organization strongly requires the ability to stop the execution and data flow fully or partially across the pipeline. This is required for pipeline hazards e.g. multi-cycle execution, data dependency, cache miss, memory access latency or false branch prediction. In such cases, the pipeline can be stalled or flushed via the pipeline controller. Such hardware schemes are commonly known as pipeline interlocking. Pipeline interlocking design is closely correlated with the compiler's instruction scheduling and data forwarding between different pipeline stages. The interlocking mechanism is applied only between the stages, where the above-mentioned hazards can occur. The following Fig. 4.8 presents such a situation. It can be observed here, that the stage 3 is accessing data memory. Due to memory access latency, the operation may take several cycles to execute. Consequently, the entire pipeline data flow needs to be stalled. In an alternative implementation, the address of the data access can be provided to the memory several stages earlier than the actual data access. This prevents the stall. However, it requires an early calculation of the memory address. To aid this, modern ASIPs often consist of a specific address generation unit [132].

Effect on rASIP Design : Similar to the issues with localized decoding, localized control affects the regularity of a re-configurable hardware block. However, in case

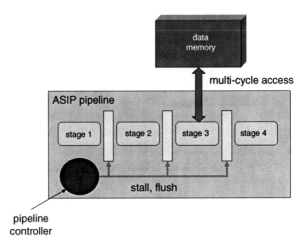

Fig. 4.8 Pipeline controller in ASIP

of a pipelined data flow within the re-configurable block, it might be necessary to contain a local pipeline controller. In order to avoid that, several design points need to be discarded for re-configurable block design. For example, memory access from a pipelined re-configurable block may not be suitable without any local control logic (ensuring the pipeline stall during to possible cache miss). In addition to that, the control signals for assessing data dependency within the re-configurable block needs to be fully derived in the base processor itself.

Special-purpose Control : Several application-specific control schemes can benefit the designer strongly. Major examples of these are zero-overhead loop stack, conditional instruction execution. *Zero-overhead loop* implements a special hardware for storing and decrementing the loop counter, thus avoiding costly processor cycles to compute this. For nested loops, a stack of such hardware can be implemented. For deeply pipelined data flow, *conditional instruction execution* offers a possibility to evaluate the condition at par with the execution. This saves the control hazard penalty originating from branch and jump instruction.

Effect on rASIP Design : The special-purpose controls can be employed in the base processor or in the FPGA part depending on which set of instructions it is targeted to. These usually results in higher routing overhead in re-configurable hardware and therefore, preferably implemented in the base processor with the control signals leading to the FPGA if required.

4.1.2 Re-Configurable Block Architecture

The design of re-configurable block evolved from early Programmable Logic Array (PLA), Programmable Array Logic (PAL) structures to the modern fine-grained and coarse-grained Field Programmable Gate Arrays (FPGAs). FPGAs, due to their higher flexibility of structure compared to other programmable logic devices, are the most common form of re-configurable hardware in use today. An FPGA consists of the following major elements.

- *Logic Element :* This refers to the basic circuitry replicated throughout to build the FPGA. The logic element can be organized in hierarchical manner, where the innermost modules are grouped to form a larger module. This larger module is referred in trade literature as Configurable Logic Block (CLB), Cluster etc. A larger module may contain heterogeneity in it by having few other circuitry for storage and/or control apart from the fundamental logic elements. The logic blocks are field-programmable i.e. they can be configured to model a wide range of functions.

  Figure 4.9 contains part of a CLB for a modern commercial FPGA [39]. As shown, the four leftmost logic elements can be configured to perform the function of a RAM, a Shift Register (SRL) or a 6-input Look-Up Table (LUT). These logic blocks can be combined via the OR-gate or can be used stand alone. To enable fast arithmetic computation, dedicated carry logic is maintained within the CLB. The results can be stored in local registers.

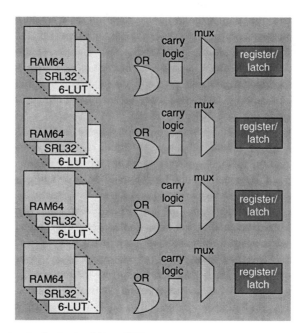

Fig. 4.9 Exemplary logic circuit within a CLB

- *Interconnect :* Interconnects are the routing resources for connecting the logic blocks in an FPGA. The interconnects are field-programmable in order to allow various possible wiring among the logic blocks. Similar to the logic blocks, the interconnects can also be arranged in a hierarchical style. The routing resources are also responsible for connecting the logic blocks to the peripheral I/O blocks. An exemplary routing structure is depicted in the following Fig. 4.10.

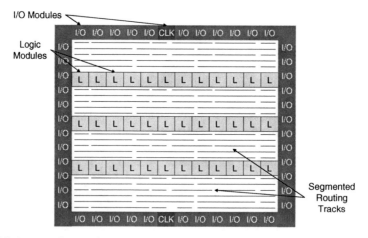

Fig. 4.10 An exemplary routing structure

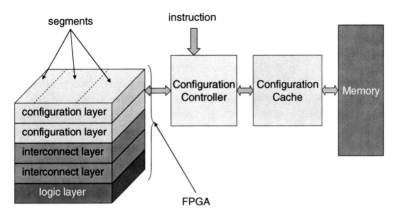

Fig. 4.11 Exemplary configuration control

- *Configuration Control :* The field-programmability of an FPGA is achieved via numerous memory bits, which set/reset connecting wires as well as configure the logic blocks. The configuration bits are usually stored in a memory location, the address of which is passed via the instruction. The configuration controller is responsible for loading the configuration bits from the memory and dispatching it to the configuration layers of the FPGA. The configuration layer can be single or multi-context. Each context is stored in a configuration layer. This is particularly useful for dynamic re-configuration. Additionally, the re-configuration time can be reduced by caching the configuration bits. For partially re-configuring the FPGA, while the other part is still executing, segmented configuration is possible. Figure 4.11 captures all the possibilities of configuration controls. The number of segments and layers in the configuration control depend on the application requirements. Apart from receiving and dispatching of the configuration bits, the configuration controller may perform further tasks. For example, the configuration controller can stall the processor execution until the current instruction's configuration is ready. This is done by checking, whether the required configuration is present in the contexts. If it is, then the context is switched to the active one. Otherwise, the configuration bit is fetched from memory/cache.

In the following, the design points introduced by different elements of an FPGA are discussed. Orthogonal to these design sub-spaces, the data flow and the control flow organization possibilities are also outlined. With the design decisions of the re-configurable block architecture, the design space of base processor are also affected. Following each design point elaboration, the effects are studied.

4.1.2.1 Logic Granularity

The granularity of logic blocks in an FPGA indicates the bit-width of the basic operators, routing wires and the class of operators. For example, a fine-grained FPGA

operates on bit-level with the fundamental logic block being an LUT. A k-input LUT
is capable of modelling any k-input boolean function. This fine-grained modelling
style is typical of commercial FPGAs. With fine granularity, it is possible to map
an arbitrary boolean function using these LUTs. This can result in less logic levels
than a boolean function implementation with a limited set of cells in the library, as
shown in [134]. Less logic levels with same logic structures result in faster implementation. However, operators with larger bit-width need to be composed of several
parallel bit-level operators. This leads to a large routing overhead, thereby causing
performance degradation. Coarse-grained FPGAs consists of wide operators and
routing resources, addressing this problem. Additionally, fine-grained FPGAs need
to be configured for every bit, thereby requiring a high volume of configuration data.
For fast dynamic re-configuration or storage of the configuration, this may lead to a
serious disadvantage. Besides that, high configuration data requires lot of switching,
increasing power dissipation.

Effect on rASIP Design : The selection of logic granularity is conceptually similar to the problem of designing an ISA. The fine-grained version offers more flexibility, at the expense of increasing configuration data and routing overhead (similar to increasing code density in RISC ISA). The coarse-grained FPGA limits the
flexibility, whereas offers faster implementation if chosen with a set of application
in mind. The granularity of FPGA logic has serious implications for the overall
rASIP. A fine-grained FPGA means almost any function can be mapped to the
FPGA. A coarse-grained FPGA limits this flexibility. The ISA of the base processor
and the custom instructions for FPGA needs to be determined according to this.
The granularity affects dynamic re-configuration overhead, which needs to be taken
into account for scheduling the application binary. The granularity also affects the
implementation efficiency of a custom instruction. This is to be considered for determining the synchronization mechanism between the base processor and the FPGA.

4.1.2.2 Interconnect Architecture

The effective usage of FPGA logic resources is reliant on the interconnect architecture. To permit high flexibility in connection, a large area of the FPGA (up to 80–
90%) is reserved for the interconnect. The design space of interconnect architecture
is extremely large, ranging from simple crossbar, mesh array, nearest neighbor connection to multi-level interconnect architectures with programmable switches and
channels. Depending on the desired flexibility and logic granularity, the interconnect architecture can have heavy (as in [39]) or light contribution to the FPGA area
(as in [135]). The granularity of the logic block contributes to the required fan-out
and fan-ins. With increasing heterogeneous input-output requirements for individual
logic blocks, an extremely flexible interconnect architecture is demanded. Usually,
such architecture is realized with channels and switch matrices. Commercial FPGAs
targeted towards a diverse set of application prototyping generally implement such
routing architecture. With coarse-grained application-specific FPGAs, local dedicated interconnects are maintained instead. These dedicated interconnects take the
form of mesh, nearest neighbor, crossbar etc. The decision is dependent on the target

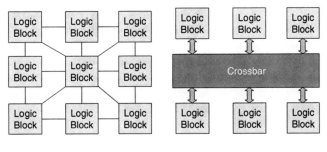

Nearest Neighbor Interconnect   Crossbar Interconnect

Fig. 4.12 Various interconnects

application area. For example, arithmetic-oriented FPGAs prefer to have nearest neighbor local interconnect schemes, as that encompasses the connection requirements of a large number of arithmetic implementations (see Fig. 4.12). A crossbar type of connection, as depicted in Fig. 4.12, offers the most dense networking possibilities in coarse-grained FPGAs. However, this comes with high implementation cost. Therefore, the number of I/O points to a crossbar is limited by the adjacent logic blocks. In case of partially re-configurable FPGAs, the interconnect wire need to have the ability to route in between segments.

To introduce controlled flexibility for the interconnecting schemes, coarse-grained FPGAs often offer a hierarchical topology. In bottom-most level of the hierarchy, the local interconnects are placed. Upper layers of hierarchy are dedicated for hopping the data across logic blocks or performing global broadcasting. One such example is shown in Fig. 4.13.

Effect on rASIP Design : As the interconnects bear the prime responsibility of passing data through the re-configurable block, they affect rASIP design apart from

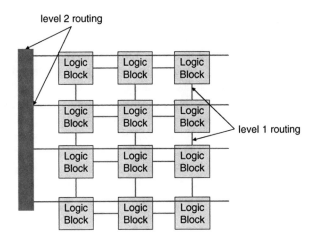

Fig. 4.13 Hierarchical interconnects

the FPGA specific design issues. The presence of global broadcasting interconnects are important for fast exchange of data in between the base processor and the FPGA. The implementation efficiency of a given application kernel is also dependent on the routing constraints [96]. A relaxed routing constraint incurs high implementation cost but, allows arbitrary control flow to be mapped on the FPGA. Therefore, the routing topology design must be done together with the targeted application domain analysis and area-performance trade-offs.

4.1.2.3 Configuration Control

By configuration control, we denote the organization of configuration layers as well as the controller responsible for loading, checking and switching configurations. For FPGAs with the possibility of partial and/or dynamic re-configuration, configuration control plays an important role in overall performance. The decision of having an FPGA to be statically re-configurable or not is a trade-off between FPGA area overhead, re-configuration time and the application performance improvement. If the FPGA is statically re-configurable then during the application run-time, no new configurations can be loaded. This leads to the situation that all possible custom instructions of the chosen application(s) must be fit within the FPGA area. To reduce the FPGA area - dynamic or partial re-configuration can be supported. However, long re-configuration time can defeat the purpose of having custom instructions at all. Such trade-offs are studied in [90].

After the basic decision to have runtime/partial re-configuration, further FPGA area can be invested to reduce the re-configuration overhead. By increasing configuration layers, employing configuration cache and enabling segmented configuration loading, the re-configuration time can be reduced. With complex configuration mechanism, a dedicated configuration controller becomes necessary. The configuration controller can also enable fast loading of configuration bits by having special interfacing with the configuration memory.

Effect on rASIP Design : The configuration control mechanism of FPGA affects the ISA of rASIP directly. For low FPGA area requirements, the custom instructions must be selected to enforce high re-use. This reduces the runtime re-configuration overhead. The instructions for loading configuration bits must also be scheduled well ahead of the custom instructions to be executed on FPGA. Partially re-configurable FPGAs allow re-configuration of some segments, while execution in other segments are ongoing. This necessitates the identification of parallel, independent custom instructions. For statically re-configurable FPGAs, the application runtime performance is to be optimized under the given FPGA area constraint. The custom instructions must fit the FPGA without overlapping. Re-usable custom instructions also bear strong advantage in this case. However, re-usable custom instructions tend to be more flexible and hence less rewarding as far as performance gain is concerned.

4.1.2.4 Data Flow

The overall data flow in the FPGA can be purely combinatorial. There the custom instruction receives the data from the base processor, performs the combinatorial processing and returns the result to the base processor. In such an FPGA, repeated loading and storing to the base processor registers might be necessary, thus creating strong pressure on the data bandwidth. The pressure can be released by local storage elements in the FPGA. The custom instructions, under such freedom, can process data independent of the base processor. The pressure of repeated memory access from the FPGA can also be released by loading the memory segments currently used to a local scratchpad memory of the FPGA. In case of coarse-grained FPGA the storage elements are often associated with each Configurable Logic Blocks (CLBs) since, the routing resources are limited and the CLB is capable of performing large computations stand-alone.

The local storages in FPGA can form a pipelined data flow, too. This allows the exploitation of temporal parallelism among successive custom instructions. More importantly, pipelining reduces the critical path significantly. As the FPGA technology naturally results in longer critical path timing than ASIC technology, pipelining in FPGA helps to bridge the timing gap. With close critical path timing, the FPGA and the base processor can run at the same clock frequency. Alternatively, the base processor clock frequency needs to be compromised for running at the same clock. Another possibility is to run the base processor and the FPGA with different clocks, leading to a multiple clock domain rASIP realization. Pipelining the FPGA data flow also introduces other issues e.g. data hazard. Hardwired solution of data hazard inside the FPGA is costly as it requires long, irregular interconnects.

Effect on rASIP Design : The base processor register file size, the data bandwidth of the interfacing and the clock frequency of the overall rASIP is to be designed taking the FPGA data flow into account. For pipelined FPGAs the instruction scheduler need to avoid data hazards or Register Update Unit (RUU) need to be maintained at the base processor.

4.1.2.5 Control Flow

Conceptually, the control flow is about transmitting the data to a data processing block. FPGA can have various degrees of control flow, offering various flexibilities. The possibilities are outlined in the following.

- Routing-dependent Control: This is typical of fine-grained FPGAs, where the data flow between CLBs and within CLBs are completely dependent on the routing resources.
- Localized Control: In coarse-grained FPGA design space, several designs have been proposed to include limited [65] or detailed control [66] in the CLBs. This is a marked shift from the data-processing oriented fine-grained FPGAs. This localized control is highly suitable for coarse-grained FPGAs. Due to wider and coarse-grained operators in the CLBs, bigger application blocks can be mapped to one CLB. Without the ability to process the control flow locally, a portion of the CLB can remain unused and/or multiple data transfers among the CLBs are required. This is again limited by typically low routing resources in coarse-grained FPGAs. However, for an application-specific FPGA, the routing can be organized to fit the control requirements. In such cases, the localized control is not required.

- Dedicated Control Region: Another possibility is to reserve a portion of the FPGA for control blocks, which drives the data-processing blocks in the rest of the FPGA [50]. Here the routing resources need not be as flexible as the fine-grained FPGAs demand but, still keeping the ability to perform local controlling. This organization is especially challenging from the perspective of multi-context and segmented configuration control.

Effect on rASIP Design : In case of data-processing oriented FPGAs, the base processor need to drive the control signals to the FPGA. These control signals need to be fixed in the interface, largely limiting the flexibility of mapping arbitrary custom instructions to the FPGA. This problem is avoided with localized control in FPGAs. With that, the issue is to leverage the control flow during selection and mapping of custom instructions.

4.1.3 Interface

The above discussions on various design points of re-configurable hardware shows that it is practically irrelevant to discuss the interfacing possibilities independently. In large number of cases the interfacing is clearly governed by the ASIP or FPGA design choices. Nevertheless, the interface decisions can be taken a priori to suit the system-level requirements e.g. memory port upper bound. Also during evolution of partially re-configurable processors, the interfacing scheme is highlighted as major design shifts. Consequently, this is taken as a basis for the processor classification in literature as well [83, 49, 48]. In the following text, a brief outline of these classifications are presented.

- *Loose Coupling:* In loose coupling, the FPGA acts as a stand-alone device. In this class, the FPGA interacts with the base processor via standard I/O interface or via co-processor access protocols. Here the data transfer between the processor and the FPGA is time consuming. Such a combination is designed for applications, where parallel blocks can be processed without much intervention. With co-processor like interfacing, the FPGA is often allowed to access memory independent of the base processor. Early partially re-configurable processors are prime examples of such coupling [35].
- *Tight Coupling:* An interface is denoted as tight when the FPGA is allowed to access internal CPU resources directly and vice versa. Tight coupling makes the task of maintaining data coherence much more difficult than loosely coupled variant. On the other hand, tightly coupled partially re-configurable processors (or rASIPs) bear the advantage of mapping arbitrary portions of the applications to the FPGA. This can be done without having to abide by stringent interface restrictions. Examples of this kind of coupling is found in several modern rASIPs [62, 85, 34].

4.2   Language-Based rASIP Modelling

The interdependent design points of rASIP make it nearly impossible to explore parts of the design space without considering the whole. There are also numerous design decisions which are difficult to parameterize. In this book, a language-based rASIP modelling is proposed. The following sections elaborate the language elements for modelling rASIP.

4.2.1   Intuitive Modelling Idea

Before starting with the language semantics, it is useful to understand the key ideas of the modelling. The complete rASIP is viewed as an Instruction Set Architecture (ISA) first. The ISA captures the syntax, encoding and the behavior of the instructions. Such a model is referred as purely behavioral model. On top of this description, we keep on adding structural information. Alternatively, the ISA is then mapped to a structural description. The structural information includes a timing model for each instruction, clocked behavior of registers and many more of such features. Structural information also includes the decision to map a part of the processor to FPGA. After these extensions of the basic model with structural information, the structure of the FPGA is modelled. The portions of rASIP, targeted for the FPGA, are then mapped to the FPGA structure. This mapping is similar to the mapping of an application to the processor. At the end of this process, two refined, structurally detailed descriptions exist. One description reflects the structure (including base processor-FPGA partitioning) and ISA of the base processor. Another description captures the structure of the FPGA. Together, the complete rASIP description is formed. This process of rASIP modelling is shown in the Fig. 4.14.

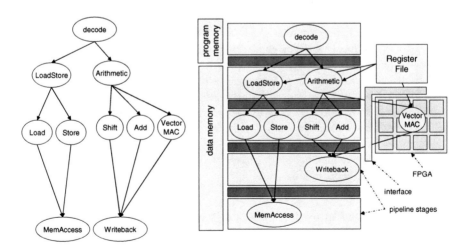

step 1 : capturing the ISA   step 2 : structuring the ISA

Fig. 4.14 Development of rASIP model

4.2.2 ISA Modelling

The aforementioned modelling concept is same as applied during processor modelling for ADLs. For this work, the ADL LISA [136] is used and extended for rASIP modelling. The basic ISA modelling remained same. For completeness, the LISA ISA modelling is described in the following.

4.2.2.1 LISA Operation DAG

The ISA comprises of syntax, encoding and behavior of processor instructions. LISA uses a modular modelling style for capturing the ISA. In LISA, each instruction is distributed over several operations. The *LISA operation* acts as a basic component, which are connected together by *activation*. The common behavior, syntax and encoding are put together in a single LISA operation. These common operations are triggered via activation from their parent operations. One common operation can be activated by multiple parent operations. Again, one operation can activate multiple children operations.

The complete structure is a Directed Acyclic Graph (DAG) $\mathcal{D} = \langle V, E \rangle$. V represents the set of LISA operations, E the graph edges as set of child-parent relations. These relations represent *activations*. For a LISA operation $\mathcal{P}$ the set of children $\mathcal{C}_P$ can be defined my $\mathcal{C}_P = \{c \mid c \in V \wedge (P, c) \in E\}$. The entire DAG can have a single or multiple LISA operations as root(s).

Group $\mathcal{G}$ of LISA operations are defined as $\mathcal{G} := \{P | P \in V\}$ such that the elements of P are mutually exclusive to each other.

4.2.2.2 Behavior Description

LISA operations' *behavior section* contains the behavior description. The behavior description of a LISA operation is based on the C programming language. By the behavior description, the combinatorial logic of the processor is implemented. In the behavior section, local variables can be declared and manipulated. The processor resources, declared in a global *resource* section, can be accessed from the behavior section as well. Similar to the C language, function calls can be made from LISA behavior section. The function can be an external C function or an internal LISA operation. In case of a function call to LISA operation, this is referred as *behavior calls*. The behavior call can be made to either a single *instance* of LISA operation or a *group* of LISA operations. Though referred as a group, the behavior call is eventually made to only one member of the group. The grouping reflects the exclusiveness of the member LISA operations.

4.2.2.3 Instruction Encoding Description

LISA operations' *coding section* is used to describe the instructions' encoding. The instruction encoding of a LISA operation is described as a sequence of several *coding* elements. Each coding element is either a terminal bit sequence with "0",

"1","don't care" bits or a nonterminal. The nonterminal coding element can point to either an *instance* of LISA operation or a *group* of LISA operations. The behavior of a LISA operation is executed only if all terminal coding bit patterns match, all non-terminal instances match and at least one member of each group matches. The root LISA operation containing a coding section is referred as the *coding root*. Special care has to be taken for the description of the coding root(s). A LISA model may have more than one coding root, e.g. for the ability to use program code with different instruction word sizes.

This set of coding roots $\mathcal{R}_c$ contains coding roots that are mutually exclusive to each other. For RISC architectures with a fixed instruction word size $\mathcal{R}_c$ contains only a single element. For VLIW architectures, each coding root $r \in \mathcal{R}_c$ decodes a set of parallel instructions.

4.2.2.4 Instruction Syntax Description

The *syntax section* of LISA operation is used to describe the instructions' assembly syntax. It is described as a sequence of several *syntax* elements. A syntax element is either a terminal character sequence with "ADD", "SUB" or a nonterminal. The nonterminal syntax element can point to either an *instance* of LISA operation or a *group* of LISA operations. The root LISA operation containing a syntax section is referred as the *syntax root*.

An exemplary LISA operation DAG with different root operations are shown in the Fig. 4.15.

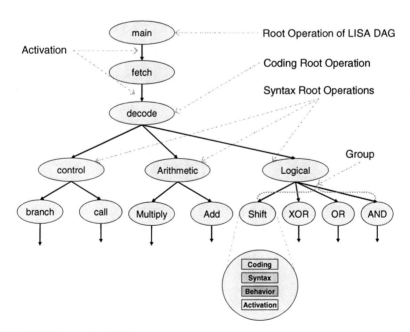

Fig. 4.15 LISA operation DAG

4.2.3 Structure Modelling : Base Processor

4.2.3.1 LISA Resources

LISA resources are a subset of general hardware resources, namely memory, registers, internal signals and external pins. Memory and registers provide storage capabilities. Signals and pins are internal and external resources without storage capabilities. LISA resources can be parameterized in terms of sign, bit-width and dimension. LISA memories can be more extensively parameterized. There the size, accessible block size, access pattern, access latency, endian-ness can be specified. LISA resources are globally accessible from any operation.

   The LISA language allows direct access to the resources multiple times in every operation. From inside the behavior section, the resources can be read and written like normal variables. Memories are accessed via a pre-defined set of interface functions. These interface functions comprise of blocking and non-blocking memory access possibilities.

4.2.3.2 LISA Pipeline

The processor pipeline in LISA can be described using the keyword PIPELINE. The stages of the pipeline are defined from left to right in the actual execution order. More than one processor pipelines can be defined. The storage elements in between two pipeline stages can be defined as the elements of PIPELINE_REGISTER. With the pipeline definition, all the LISA operations need to be assigned in a particular pipeline stage. An exemplary LISA pipeline definition is shown in the Fig. 4.16.

```
RESOURCE {
  PIPELINE pipe = {FE; DC; EX; WB };
  PIPELINE_REGISTER IN pipe
  {
      unsigned int address1;
      unsigned int address2;
      unsigned int operand1;
      unsigned int operand2;
      unsigned int result;
  }
}
```

```
OPERATION fetch in pipe.FE
{
    BEHAVIOR { .. }
    ACTIVATION { decode }
}

OPERATION decode in pipe.FE
{
    CODING { .. }
    BEHAVIOR { .. }
    ACTIVATION { .. }
}
```

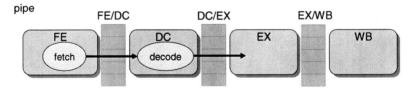

Fig. 4.16 Pipeline definition in LISA

4.2.3.3 Activations

Activations schedule the execution of child operations. In an instruction-accurate LISA model, the activations are simply triggered along the increasing depth of the LISA operation DAG. For a cycle-accurate LISA model, the activations are triggered according to the pipeline stage assignment of an operation. A LISA operation can activate operations in the same or any later pipeline stage. Activations for later pipeline stages are delayed until the originating instruction reaches the stage while operations in the same pipeline stage are executed concurrently. Activations are described in activation sections which are special parts of the description of LISA operations.

An activation section contains a list $\mathcal{A}_L$ of activation elements $\mathcal{A}_e$. All elements $\mathcal{A}_e \in \mathcal{A}_L$ are executed concurrently. $\mathcal{A}_e$ may be a group, a LISA operation or a conditional statement. For groups, only the correctly decoded LISA operation is activated. Conditional activations $\mathcal{A}_c = \langle \mathcal{A}_{if}, \mathcal{A}_{else} \rangle$ are basically IF/ELSE statements which again consist of the activation elements $\mathcal{A}_{if}$ and $\mathcal{A}_{else}$. Therefore, any level of nested conditional activation is possible.

4.2.3.4 Partitioning the Instruction-Set Architecture (ISA)

The instructions, to be mapped to the re-configurable block, need to be allocated a definite space in the overall ISA. This allocation is similar to any regular ISA branch. The speciality is attributed when this particular branch is decoded locally in the re-configurable block. Under such circumstances, the parent node of the LISA coding tree (which is resident in the base processor) needs to be aware that the following branch may incorporate new nodes even after the fabrication. This is ensured by denoting the top-most *group* of LISA operations to be mapped to FPGA, as a FULLGROUP.

4.2.3.5 Partitioning the Structure

The entire processor structure developed using the keywords so far need to be partitioned into fixed part and re-configurable part. The partitioning can be done in several levels, as listed below.

- *Set of LISA Operations :* A set of LISA operations can be grouped together to form an UNIT in LISA. This unit can be termed as RECONFIGURABLE in order to specify that this belongs outside the base processor. More than one units within one or in multiple pipeline stages can be put into the re-configurable block in this way.
- *A pipeline :* An entire pipeline structure can termed as RECONFIGURABLE and thereby moved outside the base processor. This allows modelling of custom instructions with local pipeline controlling capability.
- *Set of LISA Resources :* The special purpose and general purpose registers, residing within the base processor can be moved to the re-configurable block by setting an option to *localize register* in the LISA RTL synthesis step. By having this

option turned on, all the registers which are accessed only by the re-configurable operations (i.e. operations belonging to the re-configurable pipeline or unit) are moved to the re-configurable block. In case of units of adjacent pipeline stages being re-configurable, the pipeline registers accessed in between the units are also mapped to the re-configurable block.

- *Instruction Decoder :* The decoding of LISA operations from an instruction word is done in a distributed manner over the complete processor pipeline. A designer may want to have the decoding for re-configurable instructions to be performed in the re-configurable block itself. This can be done by setting an option to *localize decoding* in the re-configurable block. By this, the decoder of re-configurable operations is moved to the re-configurable block.

- *Clock Domains :* The base processor and the FPGA can run under different synchronized clock domains. Usually, the FPGA runs under a slower clock (or a longer critical path) for the same design implementation. The integral dividing factor of the clock is denoted by the LISA keyword LATENCY. For the custom instructions mapped onto the FPGA, the root LISA operation need to have a latency for marking a slower clock domain.

4.2.4 Structure Modelling : FPGA Description

The structure modelling [137] for FPGA blocks have three parts. First, the logic blocks. This part defines the functionality and I/O specification of an individual logic block. These logic blocks are arranged hierarchically in the topology part of the FPGA description. Finally, the interconnect part organizes the connection between the logic blocks.

4.2.4.1 Logic Block

A logic block can be written using the ELEMENT keyword. Within an element, the I/O ports are defined. For each I/O port, attributes can be specified. An attribute can be either REGISTER or BYPASS indicating that particular port can be held or bypassed while connecting. The behavior of the element is captured within BEHAVIOR section of element in form of plain C language. In order to specify a wide number of possible operators, configurable statically or dynamically, the keyword OPERATOR_LIST is used. An exemplary element definition and corresponding hardware representation can be observed in the Fig. 4.17. From the OPERATOR_LIST and the ATTRIBUTE definition, configuration bits are automatically inferred during RTL implementation. Note that, the logic blocks can be used for the purpose of routing, too. This is exemplified with the outport z in the Fig. 4.17.

4.2.4.2 Topology

The TOPOLOGY section of LISA FPGA description contains several CLUSTERs. Similar to the logic element, I/O ports and corresponding attributes can be defined

```
ELEMENT alu {
  PORT{
    IN unsigned<16> a,b;
    OUT unsigned<16> y,z;
  }
  ATTRIBUTES {
    REGISTER(y);
    BYPASS(y);
  }
  BEHAVIOR {
    OPERATOR_LIST op = {<<,>>,+};
    y = a op b;
    z = a;
  }
}
```

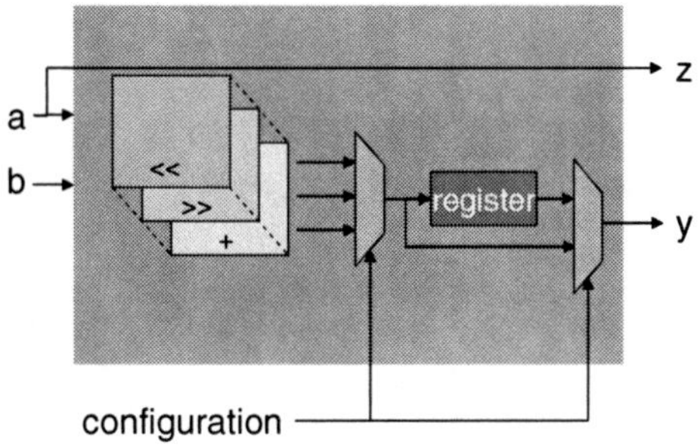

Fig. 4.17 FPGA logic block definition in LISA

```
TOPOLOGY {
  CLUSTER cluster_alu {
    PORT{ .. }
    LAYOUT{
      ROW row0 = {alu, alu};
      ROW row1 = {alu, alu};
    }
    ATTRIBUTES{ .. }
  }

  CLUSTER cluster_fpga {
    PORT{ .. }
    LAYOUT{
      ROW row0 = {cluster_alu, cluster_alu};
      ROW row1 = {cluster_alu, cluster_alu};
    }
    ATTRIBUTES{ .. }

  }
```

Fig. 4.18 FPGA logic topology definition in LISA

inside these clusters. Within the LAYOUT part of cluster, the previously defined elements can be put together in a ROW. Several rows can be then defined consecutively, building a 2-dimensional structure. A cluster can be formed using previously defined elements and/or clusters. By this process, a hierarchical structure can be formed. A completely flattened topology is nothing but a 2-dimensional structure with the basic logic elements as its nodes. An exemplary topology definition is shown in the Fig. 4.18.

4.2.4.3 Interconnect

The interconnects between the clusters and the elements can be specified in CONNECTIVITY section of the LISA FPGA description. For each cluster, a set of BASIC rules are described. Within one cluster's context several such rules connect

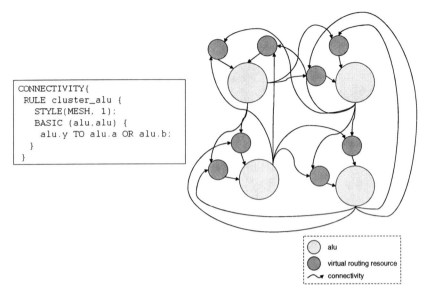

```
CONNECTIVITY{
  RULE cluster_alu {
    STYLE(MESH, 1);
    BASIC (alu,alu) {
      alu.y TO alu.a OR alu.b;
    }
  }
}
```

Fig. 4.19 FPGA connectivity definition in LISA

the I/O ports of the cluster and its children entities. Furthermore, the rules are bound by a definite *style*. Currently supported connectivity styles include mesh, point-to-point, nearest neighbor, row-wise and column-wise. In case of several rules implying multiple inputs to a single port, a multiplexer is inferred as the *virtual routing resource* for dynamic resolution. As also can be observed in the Fig. 4.19, the connectivity style is associated with a parameter to indicate the hop-length of the connections. The default parameter is 1. More on this is discussed in the following paragraph on *connectivity stride*.

Guided by this interconnect rules and style, the flattened FPGA structure takes the form of a *directed multigraph*. The graph can be defined as $\mathcal{G} = \langle V, E \rangle$. V represents the set of clusters, elements and routing resources, E represents the multiset of specified connectivity. In a completely flattened FPGA structure, V is only formed by elements and routing resources. In a non-flattened FPGA, V can contain either clusters or elements. Figure 4.19 shows an example of connectivity declaration and corresponding graph.

4.2.4.4 Connectivity Stride

To allow a more diverse set of coarse-grained FPGAs to be modelled using the proposed description style, connectivity stride is included in the interconnect description. For every connectivity style specification, an optional non-zero positive connectivity stride can be specified. The connectivity stride decides the number of hops the interconnect makes to allow a direct connection within the specified connectivity style. By this way, a mix of different connectivity styles at different hops can be established, as shown in the Fig. 4.20.

```
CONNECTIVITY{
  RULE cluster_alu {
    STYLE(NEARESTNEIGHBOUR, 1);
    STYLE(MESH, 2);

    BASIC (alu,alu) {
      alu.y TO alu.a;
    }
  }
}
```

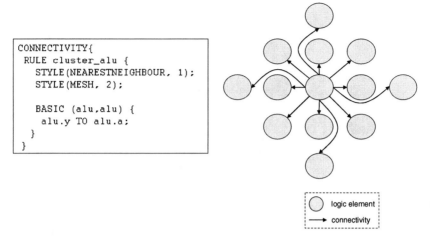

Fig. 4.20 Interconnect stride definition in LISA

A similar modelling formalism for coarse-grained FPGA have been recently proposed at [138]. However, no details of the corresponding mapping-placement and routing tools are mentioned, which makes the modelling available only for simulation purpose. The painstaking job of mapping the application on the FPGA has to be performed manually.

4.3 Language-based rASIP Design Flow

The design space exploration and implementation of rASIPs can be naturally sub-divided into two phases. The focus of these two phases are presented in the following.

Pre-fabrication Design Flow: This phase of design happens before the rASIP is fabricated. Here, the complete design space is open. The decisions involving all design sub-spaces are to be taken in this phase. Finally, the design is implemented partially on fixed and partially on re-configurable hardware.

Post-fabrication Design Flow: This phase of design happens after the rASIP is fabricated. In this phase, the base processor and the interfacing hardware is fixed. The architecture design space is limited to the possible configurations of the re-configurable block only.

4.3.1 Pre-fabrication Design Flow

The pre-fabrication rASIP design flow is shown in Fig. 4.21. The design starts from the analysis of the application(s). This analysis is done using static and dynamic profiling [113] of the application in an architecture-independent manner. The profiling

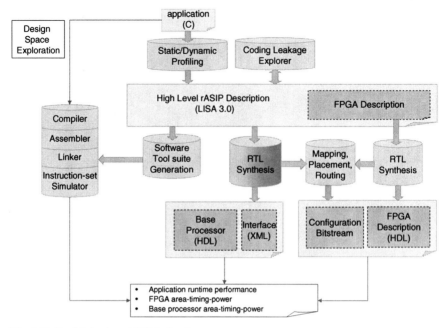

Fig. 4.21 Pre-fabrication rASIP design flow

helps to narrow down the architectural design space. With the aid of the profiler,
the decisions concerning the memory hierarchy, number of registers, processor architecture (e.g. RISC, VLIW) can be taken. The designer uses LISA language for
describing the target rASIP. The *Coding Leakage Explorer* analyzes the coding contribution of different LISA operations and determines the free coding space a.k.a
coding leakage. The higher the coding leakage in a particular branch of the LISA
operation DAG, the more number of special instructions it can accommodate. The
rASIP software tools e.g. simulator, assembler, linker, profiler are automatically
generated from the LISA description. With the aid of the LISA simulator and coding
leakage explorer, a partitioning of data flow as well as control flow is done. The
partitioning need to consider the future evolution of target application(s). Accordingly instruction space, storage space, interface bandwidth etc. must be reserved.
The LISA-based RTL synthesis tool automatically generates the RTL description of
the base processor and the re-configurable block as per the designer-directed partitioning. The partitioning generates the processor-re-configurable block interface
automatically. The complete interface is stored in the XML format. The storage
of the interface is necessary in order to ensure that the interface restrictions are
not violated during the post-fabrication rASIP enhancements. The generated HDL
description for the base processor can be further synthesized using commercial gate-
level synthesis flows [139]. The re-configurable part of the processor is internally
taken from the RTL synthesis tool and then mapped, placed and routed on to the
FPGA as per the structural specification. The output of this is a bitstream, which

is passed on to the RTL description of the FPGA for simulation. Alternatively, the re-configurable block can be synthesized to an RTL description for obtaining area-timing-power results for another gate-level/FPGA library than what is modelled in the LISA FPGA description. For simulation purposes, the LISA FPGA description is mapped to a structural RTL description via an RTL synthesis flow.

In the following the design decisions to be taken in the pre-fabrication phase are listed.

1. Overall ISA.
2. Base processor micro-architecture.
3. Re-configurable block micro-architecture.
4. rASIP data flow and partitioning.
5. rASIP control flow and partitioning.

These decisions are key to play the trade-off between the flexibility and the performance of the rASIP. For example, the availability of local decoding inside the re-configurable block allows flexibility in adding custom instructions. On the other hand, the decoder structure, being irregular in nature, adds to the interconnect cost and power budget of the re-configurable block. The decisions taken during the pre-fabrication design flow constrain the optimization benefits (or in other words, the design space) achievable during the post-fabrication design. For example, the interface between the fixed processor part and the re-configurable block is decided during the pre-fabrication flow. This decision cannot be altered once the processor is fabricated.

4.3.2 Post-fabrication Design Flow

The post-fabrication rASIP design flow is shown in Fig. 4.22. In this phase, the rASIP is already fabricated. Due to this, several parts of the design cannot be altered (labelled dark). The major design decision, which needs to be taken, is the selection and synthesis of custom instructions to reduce the application runtime. Consequently, the custom instructions need to be mapped, placed and routed on the coarse-grained FPGA.

Several custom instruction synthesis tools and algorithms have been proposed in the literature [140, 109]. In the proposed flow, the custom instruction synthesis tool (ISEGen) presented in [113] is used. This tool is able to accept generic interface constraints and produce custom instructions in LISA description format. Apart from that, manual custom instruction modelling is also possible via LISA description. During the custom instruction selection, several hard design constraints must be maintained. Those are listed below:

- The special-purpose or custom instructions must not violate the existing interface between re-configurable block and the base processor. Actually, the custom instruction synthesis tool used in the presented flow accepts this interface as an

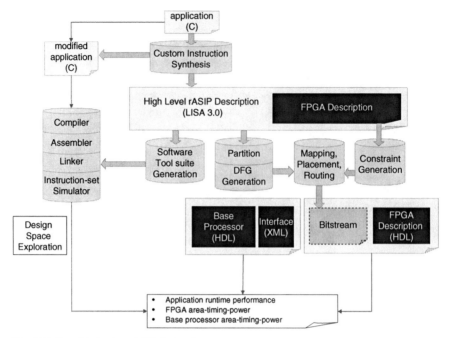

Fig. 4.22 Post-fabrication rASIP design flow

input constraint. Even after that, during the RTL synthesis phase, the fabricated interface is matched with the new generated interface.

- The additional number of instructions must not exceed the number allowed by the free coding space available.
- The overall area of the newly designed custom data path should be within the area budget of the existing re-configurable block.

In the post-fabrication design flow, ISEGen generates the custom instructions with potentially high speed-up. The calls to these custom instructions are embedded in the application. The behavior of the custom instructions are generated in form of LISA operations, which can be directly plugged in to the existing LISA model. The software tool suite for the modified rASIP description is generated automatically. The re-configurable block, as in pre-fabrication flow, is subjected to the FPGA mapping, placement and routing flow.

4.4 Synopsis

- The rASIP design space is enormous. It consists of several overlapping sub-spaces namely, base processor architecture, re-configurable block architecture and the processor-re-configurable block interface.

- The design space is captured by extending ADLs. An independent description format for the FPGA structure is presented.
- The rASIP design flow is organized into two sequential phases namely, pre-fabrication and post-fabrication phase. A language-based design flow is proposed for both phases.

Chapter 5
Pre-Fabrication Design Space Exploration

The difficulty in life is the choice.
George Moore, Novelist, 1852–1933

The continued performance benefits of the target rASIP architecture, as it will be eyeing a longer time-in-market, is largely dictated by the quality of the pre-fabrication design space exploration. Conceptually, the pre-fabrication design decisions limit the designer with the configurability and the flexibility available in the rASIP. The post-fabrication design enhancements try to excel within these preset limitations. What makes the processor design space exploration in general more difficult than, say an ASIC design, is the lack of definite parameters to quantize flexibility. For rASIPs, in addition, the configurability needs to be taken into account. These makes the task of pre-fabrication design space exploration more challenging and, when it is done with care, more rewarding.

In this chapter, a generic pre-fabrication design space exploration methodology for rASIP is proposed. The methodology is first described with the aid of a figure (see Fig. 5.1). Following that, the pre-fabrication design decisions are categorized. The application characterization is the key to the design judgements. An outline of the facets of such characterization is presented here. Later on, the LISA language support specifically for the pre-fabrication design space exploration as well as the additional exploration tools are explained in detail.

Figure 5.1 shows the pre-fabrication exploration loop for rASIP. The exploration starts from the application(s) described in a high-level language, typically C. On the basis of static and/or dynamic profiling of the application as well as the designer's knowledge the initial rASIP model is written. From the rASIP description, the software tools (e.g. Compiler, Assembler, Profiler, Instruction-set simulator) are automatically derived. These software tools are useful in providing an early estimation of the application-architecture performance. Simultaneously, a part of the rASIP description is devoted for the FPGA. A priori to that, it is necessary to partition the rASIP in the base processor and the FPGA part. This is essentially a question of future flexibility. Once important part of that is in which part of the complete ISA, the re-configurable instructions will be placed. Coding Leakage Explorer is the tool to aid the designer in that respect. Once the partitioning is done, the chosen instructions of the processor can be mapped to the FPGA description via the DFG generator and the automatic Mapping-Placement-Routing tool. As a result of these phases, a high-level estimate of the FPGA area utilization and critical path can be obtained. These profiling results can be useful to tune the FPGA description, the application or the

A. Chattopadhyay et al., *Language-driven Exploration and Implementation of Partially Re-configurable ASIPs*, DOI 10.1007/978-1-4020-9297-8_5,
© Springer Science+Business Media B.V. 2009

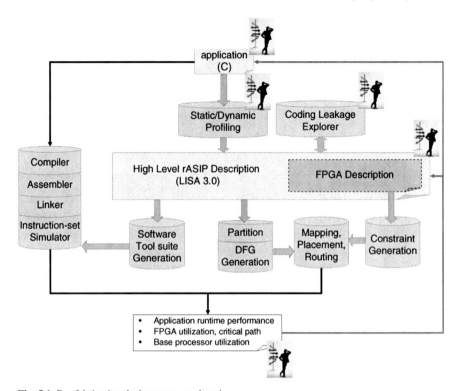

Fig. 5.1 Pre-fabrication design space exploration

base processor-FPGA partitioning. Although the FPGA exploration can be suitably done in the pre-fabrication exploration phase, due to its close connection with the pre-fabrication implementation flow the details of FPGA exploration are presented in the following chapter dealing with pre-fabrication implementation. Evidently, the implementation phase can be considered as a more detailed exploration phase, if the designer wishes so. The exploration feedback path indicated by the red line in Fig. 5.1 shows only a part of the complete exploration. For example, a more accurate estimation of the critical path can be obtained by performing HDL generation and subsequent logic synthesis tools. Of course, these can be included in a more detailed exploration loop.

Figure 5.1 shows the points where a manual intervention is necessary to interpret the results and take the course of actions for reaching the optimal design point. The complete exploration tool-suite is used for aiding an experienced designer to make the process of alteration and estimation faster.

5.1 Pre-Fabrication Design Decisions

The architectural decisions need to be taken in the pre-fabrication phase of rASIP design are enumerated in the following. In specific cases, the language extensions or tooling useful to make the decisions are mentioned.

The Base Processor Architecture The different possible (RISC, VLIW) architectures of base processor need to be weighed to select the most suitable one.

The FPGA Architecture The architectural topology, basic logic elements and interconnect structure of the FPGA needs to be decided. This is specifically aided by the language extension of FPGA description and the FPGA Mapping-Placement-Routing tool.

The rASIP Architecture The rASIP architecture organization needs to be decided, which involves extra design space than covered by the base processor architecture alone. For example, the FPGA block can be located along-the-flow or across-the-flow of the base processor architecture. In case of pipelined base processor architecture, the stage-wise synchronization with the FPGA must also be decided.

Partitioning of the Processor and the FPGA For the sake of reserving future flexibility and exploiting FPGA computational density, a part of the processor is mapped to FPGA. With the aid of LISA language element RECONFIGURABLE, this can be done.

Partitioning of the ISA between the Processor and the FPGA The custom instructions targeted for FPGA need to be allocated to one or more branches of the complete ISA. Using the LISA keyword FULLGROUP and the tool *Coding Leakage Explorer*, this can be performed.

Partitioning of the Clock Domain In case the FPGA achieves a slower clock frequency compared to the base processor clock and a multi-clock design is intended, the LISA keyword LATENCY can be used.

Storage Elements for the FPGA and the Processor This decision is crucial considering a lot of early rASIP designs have struggled to alleviate the data bandwidth issue for FPGA-processor communication. The design decisions involve storage element allocation for the base processor, storage element allocation for the FPGA as well as the organization of the data transfer between the processor and the FPGA. In cases, the communication may be unnecessary due to overlapping storage allocation.

5.2 Application Characterization

The preliminary and probably the most important phase of any architecture exploration is to develop a clear understanding of the application(s). In the following the chief characteristics of the application, which need to be looked out for, are outlined. The corresponding tools which are used in the proposed design flow are mentioned. The tools for application characterization have not been developed in the scope of current work. State-of-the-art tools are used instead. It is also important to appreciate that not all the characteristics of application are revealed by simply subjecting those to a tool. In the following, a classification is thus proposed. In any

case, the application characteristics can not be directly (or automatically) inferred
to make a design decision. This is more since, the design space is complex and
interrelated between various choices.

5.2.1 Quantifiable Characteristics

In the first category, there are are quantifiable characteristics. These characteristics
are already apparent in a host-based profiling environment. When subjected to the
target architectural settings, these characteristics are more and more evident. More
notably, these aspects of an application can be definitely quantified. As a result, the
quantifiable characteristics can be tracked early in the design flow.

- **Execution Hot-spot :** Host-based as well as target-based profiling reveals the
  applications' most frequently executed part (referred as hot-spot). The application hot-spot is usually the ideal candidate for being mapped to an accelerator
  outside the base processor. In case of rASIP, an FPGA block acts like an accelerator. For this reason, hot-spots with highly parallel, data-intensive operations
  are preferred compared to control-intensive blocks. The hot-spot can be further
  sub-divided into a block of special-purpose custom instructions. In the proposed
  pre-fabrication exploration phase, host-based profiling with *gcc gprof* [76], intermediate profiling with μProfiler [105] as well as the target-based profiling
  environment of LISA are used.
- **Memory Access Pattern :** The storage system e.g. the register file architecture
  and the memory architecture of the rASIP strongly influences the application
  performance. The data access pattern of an application can be studied beforehand
  to avoid erroneous decisions. With the aid of μProfiler [106], memory access of
  an application can be profiled. The memory access from especially the hot-spots
  can be determined to model the ideal storage element access from the FPGA.

5.2.2 Non-Quantifiable Characteristics

Non-quantifiable characteristics are more elusive to the early-stage estimation methods. In some cases, it cannot be subjected to any automated estimation at all. There,
the designer's understanding of the application and architecture can serve the best.
However, the effect of poor understanding of these non-quantifiable application
characteristics are definitely visible in the implemented design via execution speed,
power consumption and area. Another point to be noted is, that the non-quantifiable
characteristics can be hardly studied via the results of host-based profiling.

- **Programmability Requirement :**The key advantage of a processor is its programmability. This comes via the composition of the instruction-set. For an
  ASIP, the instruction-set is more specialized than say, a general purpose processor. With more specialization faster application runtime is achieved. At the

same time, this limits the ability of the processor to be adapted for a different application. It is a difficult task to identify the right set of instructions, which provides good performance over a range of applications. Considering the target set of applications decreasing for more and more specialized processing elements, the achieved performance can be conceptualized in the following Fig. 5.2. Here, the programmability is attempted to be quantified by mapping the system components across a range of applications with corresponding runtime performance benefits. For the sake of simplicity, the area and power performance are not presented. While the General Purpose Processor (GPP) delivers a basic performance over a wide range of applications, the ASIPs cover a smaller set of applications and deliver a much better performance. Therefore, it requires (as shown in the figure) larger number of ASIPs to cover the application range typically covered by a GPP.

On other hand, the FPGAs, due to its computational density advantage provides good runtime performance especially for data-driven applications. Moreover, the configurability advantage allow the same FPGA to be used for a wide range of applications, thereby making it quite general purpose in nature. Understandably, target-specific coarse-grained FPGAs fare even better in terms of runtime performance. In an effort to quantify, the flexibility aspect of processors can be directly attributed to the range of varying applications it is capable of executing. Also, the term programmability can be associated with the ease of re-programming the system component when mapping a different application on to it. This is suggested in [141]. This aspect partly depends on the modelling environment and associated tool-support. Nevertheless, to perform physical optimization of ICs, one needs to deal with much more information than to program an ASIP in high-level language. Thus, the programmability can be associated

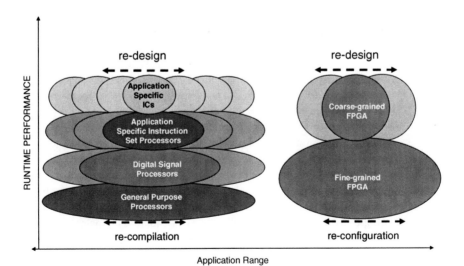

Fig. 5.2 Performance of a system component over an application range

with the complexity of input description format for a given processor/system implementation e.g. in terms of number of lines.

- **Configurability Requirement :** A characteristic, difficult to estimate and base decisions upon, is configurability. With FPGAs and rASIPs the architecture being equipped with configurability, those can sustain the performance over a range of applications by changes in the hardware as well as the ISA. Configurability can benefit the design by two ways. Firstly, the configurability can be used to deliver high performance across a range of applications. Secondly, the configurability can be reserved for future enhancements or changes of the application. By the first case, the configurability can increase the range of applications without sacrificing the performance benefits. In the second case, the configurability can ensure a longer time-in-market for the architecture. To estimate the requirement of configurability as far as the range of applications is concerned, one can continue like estimating programmability or flexibility. The configurability requirement down the evolutionary line of say, a baseband signal processing algorithm, is best envisioned by the application developer. Alternatively, the history of evolution for the same application can be studied to predict the future requirements.

- **Processor Datapath, FPGA Operator Requirement :** The custom instructions identified from the application hot-spot need to be specialized enough to give performance boost. This can be manually or automatically [112, 113] identified. The criteria for being a custom instruction can be related to the size, input-output and/or memory accesses from the *basic block* of the application. On the other hand, the specialization based on these criteria also often prevents the custom instruction to be used repeatedly for the same application or for other applications. In a typical case of engineering trade-off, the designer has to strike a balance between the performance goals and the specialization. The generality of the custom instructions can be estimated via the repeat occurrences of the same custom instruction across the application. Once the custom instruction set is determined, it can be further decomposed to identify the key operators to be implemented as FPGA basic blocks. There again, if the operators are implemented separately with the possibility of internal routing, it might increase the generality of combining those together but, at the cost of less performance. This is evident from the rise of coarse-grained FPGAs [91, 92, 41]. An approach for formally identifying the FPGA operators have been presented at [142].

- **Parallelism Requirement :**Contrary to the simple correlation of large basic block yielding lower application runtime, the application may need to have explicit parallelism for achieving the same. The call for parallelism can be block-level, instruction-level, word-level or even finer depending on the target architecture. A sophisticated target-specific C compiler is often capable of extracting a large amount of parallelism existing in the application. Typically this is done via C-level optimization techniques and dependency analysis. However, without the target architecture defined, it is difficult to identify the parallelism or to estimate how much parallelism is enough to achieve the performance goals. In most cases, the application developers are the best persons to scan through the application manually for identifying parallelism. According to that, the architecture can be designed.

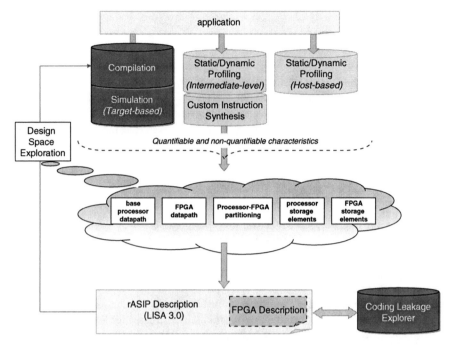

Fig. 5.3 Decision-making via application characterization

The decision-making steps in the pre-fabrication, as explained in the previous paragraphs, are shown graphically in the Fig. 5.3. The application analysis tools for host-based simulation [76] and for intermediate-level profiling, custom instruction synthesis [105, 113] are taken from state-of-the-art existing tools. For target-based simulation, rASIP-based software tool-suite is automatically derived from the rASIP model. For ISA partitioning, coding leakage explorer is developed. The tools modified or developed within the scope of this work are marked in red color. In the following sections, the algorithms and technology behind these tools are discussed.

5.3 Software Tool-Suite Generation for rASIP

The software tool s required for design space exploration of a rASIP are the following.

1. C Compiler
2. Assembler and Disassembler
3. Linker and Loader
4. Profiler
5. Instruction-set Simulator (with different levels of abstraction)

For a rASIP the ISA is still viewed as unified as a processor. For this reason, the software tools like, C-compiler, assembler, disassembler, linker, loader and profiler does not require any specific modification. The LISA-based automated software tool generation is directly used. The technology involved in these generation are elaborately covered in [10].

More specifically, the structural partitioning required for rASIP design does not influence these software tools except the Instruction-set Simulator (ISS). The required changes in the basic ISS generation flow for adapting to rASIP exploration framework are discussed in the following Section 5.3.1. The repeated partitioning of the ISA and plugging in of new composite custom instructions also require the C Compiler to be updated quite frequently. In order to reduce the design time, the compiler generation is done incrementally with the help of special features in the C compiler generator. These are discussed in the Section 5.3.2.

5.3.1 ISS Generation for rASIP

The instruction-set simulator for the processor provides a way to perform the architecture-application co-exploration without having the final processor implementation in place. From the ISA description of the processor (in LISA), the instruction-set simulator can be automatically generated as a binary. This instruction-set simulator can then run on the host machine. The application can be compiled, assembled, linked and loaded on this simulator environment. Since the application is running on a simulator environment, it is slower in comparison to the host simulation. To accelerate the simulation speed, the ISS can be run on different levels of abstraction namely, instruction-accurate, cycle-accurate and phase-accurate.

For an instruction-accurate simulator, the abstraction-level is highest as well as the speed of simulation. There, the simulator mimics the processor as a first-in-first-out buffer with every instruction entering the buffer, executing it and exiting the buffer. No two instructions can be simultaneously operating in a simple-scalar, non-VLIW mode. Clearly, the structural partitioning of rASIP does not affect the instruction-accurate ISS. For a cycle-accurate ISS, the instruction is composed of several cycles. The behavior and state of the entire processor is updated after every cycle. Therefore, more than one instruction can executing in this ISS, as in typical pipelined processors. For a phase-accurate ISS, even processor state within a cycle can be tracked. That is needed, for example, in case of interrupt execution. For both the cycle-accurate and phase-accurate mode of ISS, the structural partitioning of rASIP may dictate that a group of instructions (or their parts) are executing on a different hardware component, namely the FPGA. Under such circumstances, it might be required that the FPGA is simulating under a slower clock cycle compared to the faster clock of the base processor. Note that, this is required since the FPGA synthesis, for the same design and technology scale, yields longer critical path compared to the ASIC synthesis (applied to the base processor). A simpler solution to this problem would be to employ multi-cycle instructions, where until the completion of the instruction in one stage of the pipeline the rest of the data flow needs to be stalled. However, this is not the most general case. It is possible, for example, to have the base processor instructions executing independent of the completion of

the instructions running on the FPGA part. In a single-threaded processor special
extensions need to be done in order to have the most general case of execution. This
is what actually is attempted in the following modification.

For the ISS generation with multiple clock domains, specific extensions to the
ISS generation flow is performed. In particular we have dealt with the case, where
the FPGA clock is slow by an integral factor η compared to the base processor clock.
The factor η is denoted by *latency* in the discussion henceforth. In the following
paragraphs, first, the working principle of cycle-accurate ISS is explained. Then,
the extensions for ISS with latency is elaborated. Finally, the modifications in the
tool-flow for achieving this are outlined.

Working Principle of Cycle-accurate ISS : The cycle-by-cycle simulation of
LISA operations in an instruction is controlled by the LISA keyword `activation`.
For every operation to be executed, it needs an activation from a parent operation.
Once the operation is activated, it is scheduled to be executed. The operation is
finally executed when the execution thread comes to the pipeline stage, where the
operation is located. This is explained using the Fig. 5.4. As can be observed in
the Fig. 5.4, the instruction A is composed of a series of LISA operations termed
opA_1 to opA_5. The chain of every operation is initially triggered by the special
main operation in LISA. In the given case, opA_2 belongs to pipeline stage 2, opA_4
belongs to pipeline stage 3 and opA_5 belongs to pipeline stage 4. Although opA_2
activates both opA_4 and opA_5 simultaneously, their actual execution is delayed
until the corresponding pipeline stage. It must be remembered that the activations
are coupled with decoding, too. In this case, it is avoided to stress on the activation
issue only.

The implementation of the ISS to achieve the activation timing is explained using the Fig. 5.5. There, as can be seen, an activation data-structure is maintained
for each pipeline stage. The activation data-structure again is a set of array, with
each array element corresponding to one pipeline stage. Therefore, for n pipeline
stages we have altogether n^2 arrays. The complete activation data-structure for each
pipeline stage is shifted every clock cycle in the direction of data flow. At each

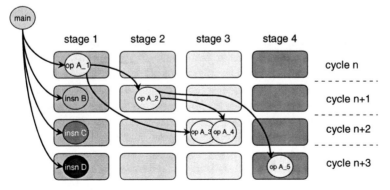

Fig. 5.4 Cycle-accurate simulation via LISA activation

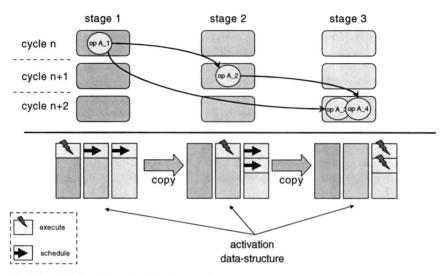

Fig. 5.5 Principle of LISA-based ISS generation

cycle, two major operations are performed on the activation data structure. Firstly, the activations emerging from the current stage is checked. All these activations are scheduled i.e. loaded in the corresponding stage of the current activation data-structure. Here, for example it is shown that the opA_1 schedules opA_2 and opA_3 in the stage 2 and stage 3 array of the activation data-structure in the n-th cycle. Now, at the beginning of the (n+1)-th cycle the whole activation data-structure from stage 1 is copied to the activation data-structure of the stage 2. At stage 2, it is found that the array corresponding to stage 2 contains an entry (basically opA_2). Thus, this entry is executed.

Extensions for Latency in Cycle-accurate ISS : The extension done to the ISS for accommodating latency is graphically shown in the Fig. 5.6. For each pipeline stage of the rASIP, an additional singular array is constructed. This array transfers the entry from the activation data-structure, when an operation with latency is encountered in the current pipeline stage of the activation data-structure. At the same time, the latency counter is decremented by 1. This additional latency data-structure is not shifted like the activation data-structure. This simply keeps on decrementing the current members until the counter becomes zero for an entry. At that moment, it is executed. There are two important points to note here. Firstly, the operations, which are activated from an operation with latency is also considered to have a latency. Therefore instead of simply getting scheduled in regular ISS, it is scheduled with the same latency as its parent operation. Secondly, with this ISS organization the operation with a latency is actually executed in a single cycle. However, the activation is delayed until a point where it reaches a zero latency count. In hardware implementation, an operation with latency demands that its input operands remain stable during the complete duration of execution. Similarly in the simulation, it is imperative for the designer to guarantee the stability of the input operands. Note that,

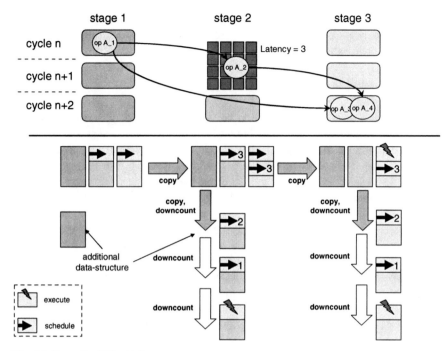

Fig. 5.6 Extending ISS with latency

unlike blocking multi-cycle operations in ISS, this organization is non-blocking in nature. This allows parallel execution of base processor operations when the FPGA-based instruction is busy due to higher latency.

Modifications in the ISS Generation Tool Flow : In the figure with ISS generation tool-flow (Fig. 5.7), the modified parts of the tool-flow are shown. The LISA language is extended with the keyword latency. Latency is declared within a LISA operation with an integer value. This modified LISA description is parsed. The

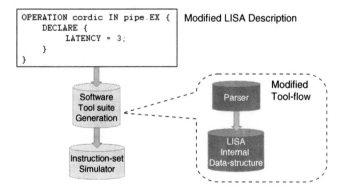

Fig. 5.7 ISS generation tool flow

pipeline stages for which operations contain latency are marked in the LISA internal data structure. Accordingly, extra array-type data-structures for storing latency are created in the generated ISS.

5.3.2 C Compiler Generation for rASIP

To enable fast design space exploration the availability of a High-Level Language (HLL) compilation flow is extremely important. It contributes to the design efficiency by firstly, allowing quick change in the application without slow, error-prone changes in the assembly code. Secondly, with the HLL compilation flow legacy software applications usually written in a high-level language can be easily targeted to the current processor. In the proposed rASIP design space exploration flow the re-targetable C compiler generator from LISA is used extensively. Since the rASIP description is an extension of LISA, the compiler generation flow can be integrated without any modifications. The re-targetable compiler generation flow from LISA is based on a modular compiler development approach. The phases of compiler (typically referred as front-end), which are machine-independent are written only once. The machine-dependent back-ends of compiler are composed of several modules as shown in the following Fig. 5.8. For a clear understanding, the tasks of the various back-end modules in a compilation flow is briefly mentioned here.

Instruction Selection The input of an instruction selector is the IR representation of the application source code. The IR representation is typically in form of a syntax tree. In the instruction selection phase the syntax tree is covered by

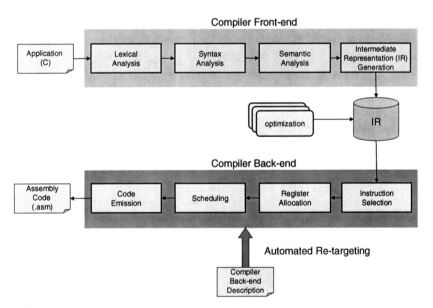

Fig. 5.8 Compiler flow

available machine instructions. Typically this is done via tree-pattern matching in a dynamic programming framework.

Register Allocation The intermediate variables used in the assembly instructions must be associated with the available physical registers of a processor. This is precisely the task of a register allocator. Typically, the register allocation is done analyzing the *liveness* analysis of each variable, building an interference graph from that and then performing bounded graph coloring [143]. In case of number of *live* variables exceeding the storage limit of physical registers, the data is temporarily stored into data memory. This event is referred as *spilling*.

Scheduling In the scheduling phase, the compiler must decide the order at which the processor instructions are to be executed. In case of multi-issue processors, the scheduler also needs to check the possibility of fine-grained instruction level parallelism and perform necessary *code compaction*. For deciding on the execution order, the scheduler checks on the data-dependency between different processor instructions as well as structural hazards. On the basis of the dependencies, the instructions can be rendered as nodes in a DAG with the dependencies as edges. The basic scheduling algorithms operate by selecting the nodes with no dependent predecessors one by one. The quality of scheduling can be strongly improved by scheduling instructions across basic blocks or by increasing the size of basic blocks by other optimization techniques.

Code Emission The code emitter simply prints the results of the previous phases in form of machine-level assembly code.

Using LISA compiler generator, the machine-dependent back-end modules can be easily re-targeted to the processor under development. As also depicted in the Fig. 5.8, the compiler back-end is automatically generated from a back-end description for a typical re-targetable compiler. In LISA, these back-end descriptions are either extracted automatically or entered manually by the designer via a Graphical User Interface (GUI).

Re-targeting the Compiler for Instructions with Latency : Here we limit our discussion within the specific features of the LISA-based compiler generator, which aids the designer to re-target the compiler after enhancing the processor with instruction-set extensions containing latency. Inquisitive reader may refer to [144, 145, 146] for a detailed study on compiler generation for the complete ISA.

After the instruction-set is extended with a custom instruction, the compiler must be extended to cover the following aspects. Firstly, the compiler back-end named instruction selector must be made aware of this new custom instruction by introducing this custom instruction to it. Secondly, the register usage of the custom instruction must be made explicit to help the register allocator distribute the register resources among the processor instructions. Finally, the dependencies of the custom instruction with the other processor instructions need to be outlined for properly scheduling the machine instructions. For re-targeting the C compiler based on new custom instructions, a complete re-generation of compiler back-end is un-necessary.

An incremental flow on top of this re-targetable framework would be fast helping the designer to speed-up the design space exploration. Exactly, this is achieved by using *inline assembly functions calls* to directly call assembly statements from the C application.

The inline assembly call defines the usage of arguments, and the custom instruction syntax itself. Within the C application, the custom instructions are called like normal function calls, which are replaced by inline assembly functions' assembly syntax portion during compilation.

An inline assembly function as well as its usage is shown in the Fig. 5.9. This function is used to perform a special operation named bit reversal in the re-configurable block. The inline assembly function is composed of two parts namely, the directives and the syntax. The arguments passed in the inline assembly call can be of definite value or a variable. For variables to be used locally within this inline assembly function, those can be declared using .scratch directive. In case it is variable, the target register resource(s) of this variable can be specified by .restrict directive. The exact syntax to be used for the inline assembly function during compilation is described following the inline assembly directives within .packs keyword. The return value of this function can be specified, as indicated, by @ directive inside the in inline assembly syntax.

Connecting the inline assembly functions to the re-targeting of compiler back-ends, it can be observed that by specifying the exact assembly syntax in the C code the necessity of extending the instruction selector is completely avoided. The inline assembly directives such as .scratch and .restrict is accounts for the back-end description used in the register allocation phase. The scheduling phase actually requires a little more elaborate information updating than only passing the directives. First of all, the inline assembly function call can be attributed with .barrier directive. This ensures that during the IR-based optimization phase, the inline-d assembly instructions are not moved across basic blocks. The delay of producing the results for instructions in the FPGA can be accounted by two ways. Firstly, the custom instruction syntax can be extended by no operation instructions (NOPs). This prevents any meaningful instruction to be issued as long as the custom instruction is executing. Of course this yields a pessimistic schedule. A more sophis-

```
asm unsigned short bitrev(unsigned short addr,
                          unsigned short n_bit_rev){
    @[
        .barrier
        .scratch temp
        .restrict temp:reg_idx<r3>
    ]

    .packs "@{temp} = bit_rev @{addr} @{n_bit_rev};",1
    .packs "@{} = @{temp};",2

}
```

Fig. 5.9 Inline assembly

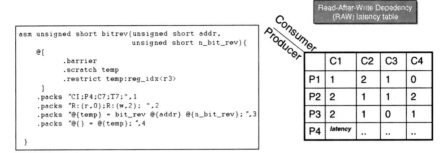

```
asm unsigned short bitrev(unsigned short addr,
                          unsigned short n_bit_rev){
    @[
            .barrier
            .scratch temp
            .restrict temp:reg_idx<r3>
    ]
    .packs "CI;P4;C7;T7;",1
    .packs "R:(r,0);R:(w,2); ",2
    .packs "@{temp} = bit_rev @{addr} @{n_bit_rev}; ",3
    .packs "@{} = @{temp}; ",4

}
```

Fig. 5.10 Inline assembly with latency

ticated approach is to update the *latency table* in the compiler back-end description. A latency table (as shown in the Fig. 5.10), a typical data-structure for generating the scheduler, contains the dependencies of various instructions (or classes of instructions). The latency table is arranged in a two-dimensional matrix. The rows of this matrix consist of all processor instruction (or classes of instructions) which writes to a resource, referred as producers. The columns of this matrix consist of all processor instruction (or classes of instructions) which reads from a resource, referred as consumers. With the new custom instructions, the latency table can be extended with new entries in the rows and/or columns. Following that, the dependencies (RAW, WAW and WAR) for each pair of instructions are to be updated in the latency table. Note that, the producer and consumer identification (P4, C7) as well as the resources produced (resource R, written at cycle 2) and consumed (resource R, read at cycle 0) by the custom instructions are specified by the .packs directive in the inline assembly function.

5.4 Partitioning the ISA : Coding Leakage Exploration

By performing coding leakage exploration, the coding contribution of different LISA operations are analyzed, explicitly marking the free coding space. This free coding space is termed as *coding leakage*. The higher the coding leakage of a particular group of instructions, the more number of special instructions it can accommodate. Thus, coding leakage plays an important role in partially re-configurable processors. Especially when the partially re-configurable processor is not defined with a fixed number of possible custom instructions (making it fully instruction-extensible), then it is important to reserve a coding branch for the custom instruction extensions such that in post-fabrication mode designer is able to model a large number of custom instructions. This is depicted in the following Fig. 5.11. In the figure, a part of coding tree can be observed. The coding contribution of the groups and the instructions are shown in binary format. From this coding tree, the coding leakage can be easily computed. Therefrom it can be deduced that if a custom instruction is added to different branches of this coding tree, the available

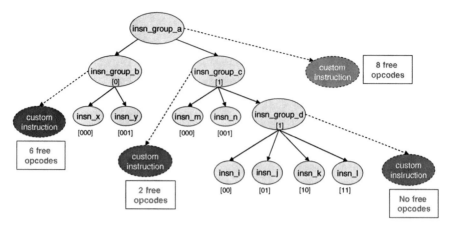

Fig. 5.11 Basic concept of coding leakage

number of possible encodings differ. The highest coding availability is obviously at the root of this simple coding tree with 8 free op-codes. In practice, however, the designer may find it more suitable to group the custom instructions with similar operations. That allows to organize the data flow among the chain of operations as well as across the pipeline stages. By adding the custom instructions to the similar operations' group, the effort to re-target the compiler can also be minimized. On the other hand, for complex coding tree it is extremely difficult to scrutinize every operation's coding contribution and determine free op-codes in a given branch. The stand-alone tool named *coding leakage explorer* does exactly that. In the following paragraphs, the working principle and the algorithms involved in the tool are discussed.

Instruction Grammar File : As a starting point, the *coding leakage explorer* needs to obtain the detailed coding hierarchy of the ISA. That is automatically generated by parsing the LISA description and stored in form of an instruction grammar file. The grammar file represents the instruction set in the Backus-Naur Form (also referred as PaniniBackus Form). Table 5.1 shows an exemplary instruction grammar. For this example, the instruction word width is 16 bit and there are 16 available registers indexed by src_reg and dst_reg. Don't cares are represented by the symbol x. Note that, this instruction grammar is essentially an instruction coding tree represented in another form.

Formally, the instruction grammar can be defined as following.

- Terminal (T): It is either '0' or '1' or 'x' (don't care).
- Non-Terminal (NT): Syntactic variables that denote the sets of strings. (containing terminals and/or non-terminals).
- Production (S) : $P : \alpha_1\alpha_2...\alpha_n$, where $\alpha_i \in NT \setminus \{P\}$ or $\alpha_i \in T$, $P \in NT$, $\forall i \in N$, where N is set of positive integers. In this context, we define P as the *Producer*.

Table 5.1 Exemplary instruction grammar

insn	: add dst_reg src_reg src_reg ‖ sub dst_reg src_reg src_reg
	‖ ld dst_reg imm ‖ jmp cond_reg imm
	‖ nop
add	: 0001
sub	: 0010
jmp	: 0011
ld	: 10
src_reg	: xxxx
dst_reg	: xxxx
cond_reg	: xx
imm	: xx xxxx xxxx
nop	: 01xx 0000 0000 xxxx

- Rule: A set of productions (S) having the same producer P. Formally, $S = \{S_i \mid P_i = P, \forall i \in N\}$,

  where $S_i \rightarrow P_i : \alpha_1 \alpha_2 ... \alpha_n$

With the grammar file, the coding hierarchy of the LISA operations is explicitly presented. For each instruction containing a chain of LISA operations the information about op-codes and operands including their position and bit-width can be obtained.

Data-structure for Coding Leakage Exploration : The algorithm for determining coding leakage requires an efficient data-structure for storing the coding tree. The coding data-structure is a DAG with the nodes indicating the non-terminals and terminals of the instruction grammar. The edges of the DAG can be of two type e.g. vertical edges (v_egdes) and horizontal edges (h_edges). Both kinds of edges are associated with the producer non-terminals and their rules. From a producer non-terminal, the concatenated productions are connected via h_edges. The productions, which are separated by | symbol thereby indicating alternative possible rules, are connected with the producer via v_edges. While there can be multiple vertical children of a given node, there can be only one horizontal child for each of it. The data-structure is populated by parsing the instruction grammar file.

Coding Leakage Representation : Irreducible Form : The free coding space i.e. leakage is all possible combinations other than the encoding bits which are assigned to the existing LISA operations. This can be obtained in a store and compare fashion. Lets suppose operation A has opcode 0000 and its the only opcode used in this 4 bit location of instruction. Therefore all the possible available coding combinations i.e. 0001, 0010, 0011, ..., 1111 are the leakage in that local branch (GroupA → GroupB → OpA). This *local leakage* is then compared with other branches i.e (GroupA → GroupB → OpB),(GroupA → GroupC → OpC) and (GroupA → GroupC → OpD) to check either they are actually free coding leakage or not. This leakage is initially determined for each branch of the instruction tree and then propagated over the complete instruction set. This procedure is shown in the Fig. 5.12.

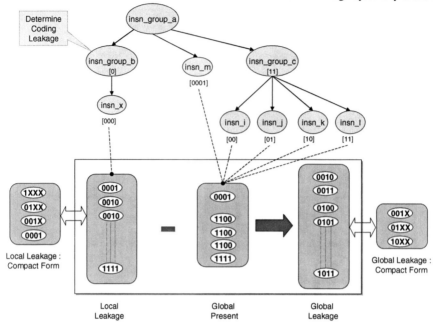

Fig. 5.12 Determination of coding leakage and irreducible form representation

As it can be observed here that if there are n bits allocated for the opcode of a certain operation in LISA, there can be maximum $(2^n - 1)$ possibilities of free coding space. This huge number of possibilities need to be matched against other coding branches to obtain the final coding leakage. To improve the storage and execution efficiency of coding leakage exploration, a compact form of expressing the leakage is introduced. This is shown in the Fig. 5.12. There, in addition to 0 and 1 bits of a coding sequence, don't care bits are also used. The algorithm to obtain the compact form representation from a given coding sequence is straightforward. First, one needs to invert the first bit of a given sequence followed by all don't cares. Then, the second bit of the original sequence needs to be inverted while first bit remains same as in the original. These two bits are to be followed by don't cares. Similarly, this process can be repeated by maintaining the original bits of a given sequence, inverting the original bit at one location followed by don't cares until the last bit is inverted to produce a new sequence. With this form of representation, only n bit-vectors are required instead of $(2^n - 1)$. The data-structure used for such representation is simply a list of strings, referred as *Leakage* henceforth.

5.4.1 Coding Leakage Determination Algorithm

The pseudo-code of the algorithm for determining coding leakage on the basis of the above-mentioned data-structure is given in the algorithm 5.1. The input to the function *Find_Coding_Leakage* is a particular node, which marks the beginning of a branch. Using this function, the leakage for the coding branch below this node is obtained. Leakage is defined here as a list of coding strings. Each of these strings consists of an array of elements from the set 0, 1, x. The function first calls each vertical children of the current node, obtains their corresponding leakages and concatenates those leakages. The same method is repeated for the horizontal children of the current node. The concatenation of leakages for vertical and horizontal children are treated in unique ways, which are discussed in the following paragraphs. Note that, this is not the leakage of the node when the parent and sibling nodes of this branch are taken into account. For obtaining that, first the coding leakage for the *root* node of a DAG is determined. Following that, the leakage is propagated below to each branch of the DAG. During the leakage propagation, the local leakage of a node is compared against the global leakage of the root node. The elements of local leakage, which are also present in the global leakage are marked as global leakage for the current node. Alternatively, this can be termed as the global leakage within the scope of the current node. From a designer's perspective, this is the leakage of a node.

Concatenation of Coding Leakage for Horizontal Siblings : The concatenation of coding leakage of horizontal siblings of a given node is straightforward since the nodes involved in the function are working on disjoint locations of coding. The algorithm first determines the *present coding* by subtracting leakage coding from all possible codings within the scope of the current node. In practice, a parallel data-structure is maintained for present coding, which is propagated along with the coding leakage to save computation time. Once this is performed for both of the nodes, first the leakage coding of both the nodes are joined by plain string concatenation. Then the alternate present and leakage combinations are joined by string concatenation to be appended to the overall coding leakage. The pseudo algorithm is presented in algorithm 5.2. Note that, the present coding as well as leakage coding are stored in the irreducible form.

Algorithm 5.1: Pseudo Code of the Function Find_Coding_Leakage

---

Input: *node*
Result: *Leakage*
begin

    Leakage *curr_Leakage* $= \emptyset$;
    // Leakage Concatenation for Vertical Children
    $vchild_{node} = node \rightarrow Next\_Vertical\_Child$;
    while $vchild_{node}$ **do**
        | Concatenate_Vertical(curr_Leakage, Find_Coding_Leakage($vchild_{node}$));
        | $vchild_{node} = node \rightarrow Next\_Vertical\_Child$;

    // Leakage Concatenation for Horizontal Children
    $hchild_{node} = node \rightarrow Next\_Horizontal\_Child$;
    while $hchild_{node}$ **do**
        | Concatenate_Horizontal(curr_Leakage, Find_Coding_Leakage($hchild_{node}$));
        | $hchild_{node} = hchild_{node} \rightarrow Next\_Vertical\_Child$;

end

Algorithm 5.2: Pseudo Code of the Function Concatenate_Horizontal

---

Input: $curr\_Leakage$, $hchild_{node}\_Leakage$
Result: $curr\_Leakage$
begin

  $Leakage\ curr\_Present = \overline{curr\_Leakage}$;
  $Leakage\ hchild_{node}\_Present = \overline{hchild_{node}\_Leakage}$;
  $curr\_Leakage = \emptyset$;
  $curr\_Leakage$.append($curr\_Leakage$.concat($hchild_{node}\_Leakage$));
  $curr\_Leakage$.append($curr\_Present$.concat($hchild_{node}\_Leakage$));
  $curr\_Leakage$.append($curr\_Leakage$.concat($hchild_{node}\_Present$));

end

Concatenation of Coding Leakage for Vertical Siblings : The concatenation of coding leakage for vertical siblings of a given node is performed by repeatedly calling the function Concatenate_Vertical, of which the pseudo algorithm is given here (algorithm 5.3). As specified in the algorithm, for the first run of it, the coding leakage is obtained by getting that of the first vertical child. This actually initiates a depth-first traversal until a terminal node is met. For the subsequent runs, the present coding of the vertical sibling is subtracted from the coding leakage gathered so far. Thus by repeatedly cleaning up the coding leakage taking vertical siblings into consideration the global coding leakage within the scope of current node is obtained. The subtraction of present coding from the coding leakage is done in two steps. In the first step, the irreducible form of coding leakage is compared with the present coding of the given vertical sibling. Sometimes for performing comparison, the irreducible form needs to be expanded by replacing x by 0, 1. This is done in the first step. In the second step, the actual comparison is done and present codings from the sibling node is removed. It is important to note that, the coding leakage from the vertical sibling need not be considered. This is already covered by the initial coding leakage of first vertical child as all vertical children function on the same coding location.

The complete coding leakage determination procedure can be performed via a Graphical User Interface (GUI). In addition to the determination of node-specific

---

Algorithm 5.3: Pseudo Code of the Function Concatenate_Vertical

---

Input: $curr\_Leakage$, $vchild_{node}\_Leakage$
Result: $curr\_Leakage$
begin

  if $curr\_Leakage == \emptyset$ **then**
  $\quad\lfloor\ curr\_Leakage = vchild_{node}\_Leakage$;

  else

  $\quad\quad Leakage\ vchild_{node}\_Present = \overline{vchild_{node}\_Leakage}$;
  $\quad\quad$ // check if current leakage needs expansion
  $\quad\quad$ Compare_And_Expand($curr\_Leakage$, $vchild_{node}\_Present$);
  $\quad\quad$ // remove present codings of sibling node
  $\quad\quad\lfloor$ Compare_And_Remove($curr\_Leakage$, $vchild_{node}\_Present$);

end

---

coding leakage, the user obtains additional information about the position of the coding bits in complete instruction word for a given node. The designer also gets the information about the pipeline stage in which a given node belongs. With these tools, it is possible to check the coding leakage of various instruction branches. Consequently, those encoding bits can be reserved for custom instructions and the coding leakage can be studied again.

5.5 Synopsis

- The pre-fabrication exploration begins from identifying the application characteristics and modelling the processor according to that.
- Traditional processor design tools need to be tailored for modelling rASIPs. The adaptation for ISS generation is presented. The specific feature of retargetable C compiler, used for this work, is elaborated.
- To identify the proper placeholder for the re-configurable instructions in the rASIP ISA, a GUI-based tool is developed. The tool identifies unused opcode bits in different branches of the ISA coding tree.

Chapter 6
Pre-Fabrication Design Implementation

Form follows function.
Louis Henri Sullivan, Architect, 1856–1924

Pre-fabrication design implementation phase begins when the designer is satisfied with the outcome of the pre-fabrication exploration (previous chapter) results. The results are procured by various tools e.g. the instruction-set simulator, the FPGA mapper. The results indicate the performance of the application(s) when mapped to the target rASIP. Surely, these results are obtained at a higher level of abstraction than the physical implementation level. For this reason, the designer can perform exploration in a lower level of abstraction such as Register Transfer Level (RTL). Thus, it must be noted, that the design implementation phase elaborated in the current chapter does not preclude design exploration. It is possible to use more physically accurate results in order to re-design the processor or to re-write the application. As in typical cost-accuracy trade-offs, the increased accuracy of design exploration elongates the exploration phase itself. The RTL description marks the beginning of most standard commercial tools [139, 147, 148], which allow an optimized flow till the fabrication. Therefore, it is till the generation of RTL description, for which the tools and algorithms are developed in this book.

In the design implementation phase the rASIP requires several significant extensions compared to traditional processor design flow. To do away with the arduous job of manual RTL description for the complete processor, ADLs have proposed (semi-)automatic RTL code generation for quite some time [149, 150, 151, 152, 16]. The ADL LISA, used in the current work, not only contains a strong background of RTL code generation, but also showed that the automatically generated RTL code can outperform manually developed models by incorporating high-level optimization techniques [153, 154, 155, 156].

In the current chapter, the basic LISA-based RTL code generation is first explained. Following that, the extension in the RTL code generation for partitioning the processor into a fixed part and a re-configurable part is elaborated. The structural part in the LISA description consists of detailed FPGA specification, which is synthesized to a structural RTL description. The details of this FPGA synthesis process are mentioned in this chapter. Finally, the re-configurable part of the processor needs to be placed and routed on the FPGA. In the FPGA mapping process, current state-of-the-art placement and routing algorithms from the domain of fine-grained FPGAs are used. The algorithmic interfaces are modified to support a widely varying coarse-grained FPGA structural descriptions and to allow a RTL simulation framework

A. Chattopadhyay et al., *Language-driven Exploration and Implementation of Partially Re-configurable ASIPs*, DOI 10.1007/978-1-4020-9297-8_6, © Springer Science+Business Media B.V. 2009

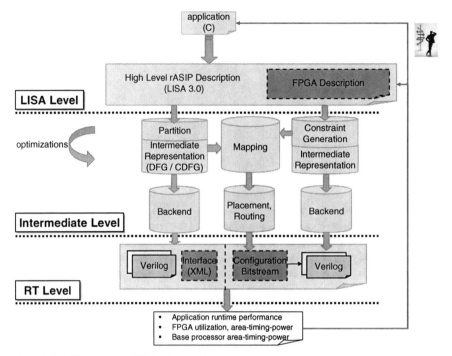

Fig. 6.1 Pre-fabrication rASIP implementation

with the FPGA description captured at RTL. The entire process of pre-fabrication
rASIP implementation is synoptically captured in the following Fig. 6.1. As shown
in the figure, the Intermediate Representation (IR) of the complete processor is partitioned before proceeding to the back-end. The re-configurable part is passed on to
the mapping phase and the fixed part is processed by the back-end to produce RTL
description in VHDL and Verilog. It must be noted that, it is also possible to pass the
entire processor via the back-end and generate a partitioned RTL description. There
onwards, the designer can process the re-configurable part's RTL description using a
commercial FPGA synthesis flow. The design flow presented in the Fig. 6.1 merely
suggests for the most inclusive implementation, where the FPGA itself is also designed. As shown in the figure that it is possible, and often beneficial, to perform a
more detailed design space exploration by obtaining the results from RTL-to-gate
synthesis and RTL simulation.

6.1 Base Processor Implementation

The process of generating the base processor from the ADL description can be
divided into two phases namely, the construction of the IR and the execution of
language-specific back-ends. The optimization engines, too, play an important role

in generating efficient RTL description. Still, to maintain the focus of the current work the IR definition, construction and the back-ends are only discussed here. Interested readers may refer to the [157] for a detailed discussion on the optimization techniques employed.

6.1.1 IR Definition

The definition of the IR is aimed at combining benefits of high-level ADL description with the detailed information available at RTL. In this section, the basic IR elements are first outlined. Following that, the semantical information annotated to each of these elements are explained.

Here, the IR elements are defined formally.

IR-process Each IR-process encapsulates hardware behavior, which is represented by either a Control Data Flow Graph (CDFG) or a Data Flow Graph (DFG). $\mathcal{A}$ denotes the set of processes $\mathcal{A} = A_1, A_2, ..., A_{N_A}$, which is used to model the whole target architecture. The processes provide a functional abstraction of the architecture on RTL. Given, the similarity of process functionality can be used to group them into even more generic functional blocks, such as decoder or pipeline controller.

IR-unit IR-unit, an additional abstraction layer, is used for conveniently grouping one or more IR-processes of similar functionality. A unit U_i is defined as set of processes $U_i \subset \mathcal{A}$ with $\cup_{i=1}^{N_u} U_i = \mathcal{A}$ and $\forall U_i, U_j \in \mathcal{U}$ and $i \neq j$ - $U_i \cap U_j = \emptyset$. $\mathcal{U}$ denotes the set of all IR-units.

IR-signal An IR-signal is a connection between processes and dedicated for communication. Information between the processes is exchanged via these IR-signals. $\mathcal{S}$ denotes the set of signals required to model the target architecture, with $\mathcal{S} = S_1, S_2, ..., S_{N_S}$.

IR-path IR-signals required for exchange of information are grouped together in an IR-path. An IR-path P_i is defined as a set of signals $P_i \subset \mathcal{S}$ with $\cup_{i=1}^{N_p} P_i = \mathcal{S}$ and $\forall P_i, P_j \in \mathcal{P}$ and $i \neq j$ - $P_i \cap P_j = \emptyset$. $\mathcal{P}$ denotes the set of all paths. For example, a simple assignment to a register array in LISA such as R[address]=data, is represented in an HDL by three different signals. Those comprise data, address and enable flag indicating the validity of the data and address values. An IR-path groups those signals modelling a particular transaction.

A functional abstraction of the architecture is defined by a multi-graph.

$\mathcal{G}_{IR}$: $\mathcal{G}_{IR}$ is a directed multi-graph $\mathcal{G}_{IR} = (\mathcal{U}, \mathcal{P})$. Here, the vertex set is $\mathcal{U}$ and the edge set is $\mathcal{P}$. Typically, at the RTL abstraction, a hierarchical model structure is provided by nested entities in VHDL (modules in Verilog). Following this principle, the IR employs IR-entities to encapsulate the functionality provided by IR-units. An IR-entity E_i is defined as set of IR-units with $\cup_{i=1}^{N_E} E_i = \mathcal{U}$. Finally, a hierarchical abstraction of the architecture is defined by a tree $\mathcal{T}_{IR}$. $\mathcal{T}_{IR}$ is defined as a directed tree $\mathcal{T}_{IR} = (\mathcal{E}, \mathcal{I})$. Here, the vertex set is $\mathcal{E} = E_1, E_2, ..., E_{N_E}$. $\mathcal{I}$ denotes the set of edges. A directed edge $\langle E_i, E_j \rangle$ represents the relation between E_i and E_j as in a parent-child relation of HDL component instantiation.

IR Definition: The overall IR is defined by the functional abstraction $\mathcal{G}_{IR}$ and the hierarchical abstraction $\mathcal{T}_{IR}$.

An example of the IR is depicted in Fig. 6.2 including entities, edges(dotted arrows), units, paths (solid lines) and the processes included in an unit. To further classify the IR elements, to make the data-structure easily operable, specific types of IR-units and IR-paths are derived. This is done by attaching special semantical information to these elements. Table 6.1 contains an exemplary list of few types of IR elements. The types of entities, units and paths have been chosen by implementing and evaluating several existing architectures. For example, the pattern matcher unit is dedicated to recognize particular instruction bit pattern and to set the coding path accordingly. The information provided by the coding path as well as run-time conditions are evaluated by the decoder unit to set the correct control signals via the activation path. Those signals are, for example, utilized by the data path units. Understandably, the IR-elements' semantic types can be extended with new types.

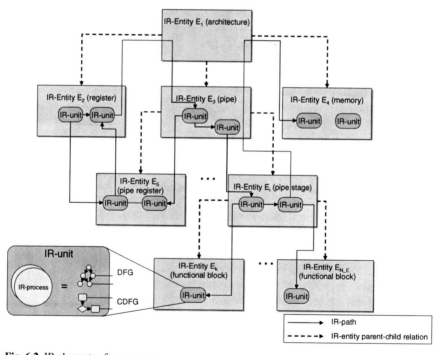

Fig. 6.2 IR elements of a processor

Table 6.1 List of semantically annotated IR elements

IR-entity	IR-unit	IR-path
testbench	data_path	register_resource
architecture	reconfigurable data_path	memory_resource
register	register	signal_resource
memory	pipe_register	coding
pipeline	memory	activation
pipeline_stage	simulation_memory	pipe_reg_control_stall
pipeline_register	signal	pipe_reg_control_flush
functional_block	pipe_controller	clock
	clock	reset
	decoder	
	pattern_matcher	
	architecture	
	pipe_controller	
	reset	
	pipeline	

6.1.2 IR Construction

The ADL-specific frontend constructs the IR from the information given in the ADL. Pieces of information, which are directly extracted from the ADL model, are called explicit information. For maintaining high abstraction, ADLs do not cover all the hardware details, as a fast architecture exploration is the primary goal. Therefore, the missing information must be derived from common knowledge about processor design. The pieces of information introduced in this manner are called implicit information. The proposed IR supports the usage of explicit and implicit information in the hierarchical representation $\mathcal{T}_{\mathcal{IR}}$ as well as in the functional representation $\mathcal{G}_{\mathcal{IR}}$. For example, the LISA frontend uses a hierarchy template which instantiates the registers and memories entities within the architecture entity (see Fig. 6.2). This fact allows various optimizations, as different templates may coexist, each for a particular optimization. The construction of IR happens in the following phases as discussed below.

Structuring In the first step, called structuring, the directed tree $\mathcal{T}_{\mathcal{IR}}$ is instantiated. The different IR-entity types, as from conventional processor design knowledge, are instantiated hierarchically by following the architecture described using LISA. The entity structure is equivalent to the structure of the final hardware model on RTL generated by the back-ends.

Mapping In the second step, the information given in the ADL is mapped onto the functional representation of the hardware $\mathcal{G}_{\mathcal{IR}}$. For example, information about resources is used to map the resource-units while information about the instruction set is mapped onto the decoder-units. However, during this phase only a path subset $\mathcal{P}_{\mathcal{M}} \subset \mathcal{P}$ is created. The remaining paths $\mathcal{P} \setminus \mathcal{P}_{\mathcal{M}}$ are built up in the next phase.

CDFG/DFG Creation Every IR-process must be implemented by a CDFG or a DFG. This is done by an extended C-parser. The parser maps the behavior code inside LISA operations, which is represented in C language, to the CDFG representation. Due to the relatively straightforward mapping to the CDFG format, it is used as the default data-structure. Later on for optimization purposes and for FPGA library mapping, placement and routing, the CDFG is suitably converted to DFG. In this book, DFG representation is used extensively. The details of CDFG to DFG conversion are presented in due course of its usage.

6.1.3 Back-End for HDL Generation

As discussed earlier, the processes and signals required in a HDL are embedded within the basic elements of the IR. To extract the hardware description of the target architecture, the cover of IR-units and IR-paths must be dropped to expose processes and signals. This is the first time in the RTL processor synthesis flow, that the semantical information is omitted. The consequent mapping is already evident. An IR-entity with one or more IR-unit (U_i) is mapped to an *entity* in VHDL and a *module* in Verilog. The IR-processes, previously covered within an IR-unit U_i, are mapped to *process*es in VHDL and *always blocks* in Verilog. The IR-signals, grouped in an IR-path P_i, generates *signal*s and *port*s in VHDL and *wire*s and *reg*s in Verilog. When generating a specific HDL, the language-specific requirements of various HDLs need to be considered. One of the prime constraints is the type-propagation. The Verilog HDL is loosely typed, whereas in VHDL the data type is imposed strongly. To cope with this constraint, a specific type-propagation function is inserted before generating a particular HDL. In addition to the RTL description in various languages, the scripts for driving the RTL simulation and gate-level synthesis are generated automatically for VHDL and Verilog.

6.2 Partitioning the Structure : Specification Aspects

In this section, the partitioning of rASIP structure in a fixed block and a re-configurable block is explained. Firstly, the language extensions are studied followed by the actual implementation part. The implementation aspects rely fundamentally on the traditional processor implementation using IR, explained in the previous section.

6.2.1 LISA Language Extensions for Partitioning : Recapitulation

The intricacies of LISA language extensions for partitioning the rASIP between a processor and a re-configurable block are explained earlier (refer Chapter 4). Here, a brief recapitulation is presented for continuous reading. The language syntax allows parts of the data-path, resources and control-path to be targeted towards the re-configurable block. The exact language syntaxes and their implications are pointed in the following accompanied with a pictorial overview.

1. In LISA, the *operations* are used to represent parts of an instruction. A set of LISA operations can be grouped together to form an UNIT in LISA. This unit can be termed as RECONFIGURABLE resulting in a partition. Note that, the partition is visible only after the synthesis of LISA description to RTL. It allows the high-level designer to concentrate on the overall processor design with the re-configurable block being a part of say, a pipeline stage. This is shown in the following Fig. 6.3.
2. An entire pipeline can be termed as RECONFIGURABLE and mapped to the re-configurable block (Fig. 6.4).
3. The resources of the processor can be partitioned to belong to the re-configurable block if firstly, those are accessed solely from the data-path already assigned to the re-configurable block and secondly, an option called *localize register* is set during RTL synthesis. Such a case is graphically shown in the Fig. 6.5.
4. The control-path involving the instructions (or parts of those), which are already assigned to the re-configurable block, can be partitioned to belong to the re-configurable block, too. This is done by setting an option called *localize decoding* during RTL synthesis. Decoder can be localized with or without the option to allow further new opcode to be incorporated into the processor ISA. This can be done via the LISA keyword FULLGROUP.

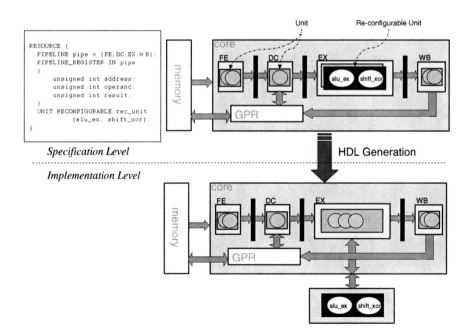

Fig. 6.3 Re-configurable unit in LISA

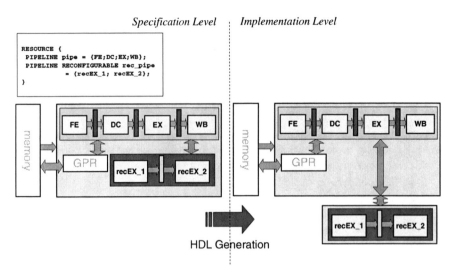

Fig. 6.4 Re-configurable pipeline in LISA

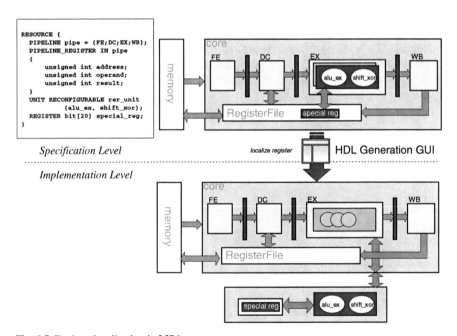

Fig. 6.5 Register localization in LISA

5. The base processor and the FPGA can run under different and synchronized clock domains. Usually, the FPGA runs under a slower clock (or a longer critical path) for the same design implementation. The integral dividing factor of the clock is denoted by the LISA keyword LATENCY.

6.3 Partitioning the Structure : Implementation Aspects

On the basis of aforementioned specification extensions, the LISA-based *HDL Generator* is extended. The prime contributor to this extension is the IR-based data-structure, which is implemented in a flexible manner. Using this IR, an algorithm for *unit movement* is implemented. Powered with this unit movement algorithm, it is possible to partition the structure in different degrees. In the following sub-section, this basic algorithm for unit movement is discussed at first. On that basis, the implementation of structural partitions are explained.

6.3.1 Unit Movement

For internal usage, IR-units are conveniently used as a wrapper around the IR-processes. Thus, the IR-units can be considered as the singular constituent of an IR-entity. IR-units are characterized by two special member elements. Firstly, the set of IR-processes. Secondly, the set of IR-paths connecting the internal IR-processes to the external IR-entities and so on. Clearly, with the unit movement, these two members need to be adjusted. Thanks to the data-structure organization, the IR-processes only interact with the wrapper IR-unit along with the IR-unit's member IR-paths. This reduces the task of unit movement purely to the re-adjustment of the IR-paths. The algorithm is explained in the following paragraphs in bottom-up fashion. Initially, the movement of IR-unit between one-level of entity hierarchy is explained. On top of that, the overall unit movement algorithm is discussed.

The one-level unit movement is outlined in the previous algorithm 6.1. The algorithm is implemented within the scope of an IR-entity. The IR-path structure is internally built up of path segments consisting of *path-ports*, *path-signals* and *path-endpoints*. Path-endpoints are those connected to the IR-units. Path-ports mark the connection to an IR-entity. Path-signals connect between path-ports and path-endpoints. At first, the IR-unit to be moved is removed from the source entity's member list and appended to the list of member IR-units of the target entity. Then, for each path-endpoint of the IR-unit under movement, the nearest path-port or

---

Algorithm 6.1: MoveUnitOneLevel

Input: *unit, source_entity, target_entity*
begin
    source_entity → removeUnitFromList(*unit*);
    target_entity → addUnitToList(*unit*);
    bool moveup = (*source_entity* ∈ *target_entity*);
    foreach *pe* ∈ *unit* **do**
        IR_Path pe_p = *pe* → getNextPortOrEndpointAwayFromUnit();
        IR_Path pe_s = *pe* → getNextSignalAwayFromUnit();
        if *moveup* == *true* **then**
            if *pe_p* → getEntity() == *source_entity* **then** *pe* → removePathSegment(*pe_p, pe_s*);
            else *pe_s* → insertPathSegment(*source_entity*);
        else
            if *pe_p* → getEntity() == *target_entity* **then** *pe* → removePathSegment(*pe_p, pe_s*);
            else *pe_s* → insertPathSegment(*target_entity*);
end

Algorithm 6.2: MoveUnit

Input: *unit, source_entity, target_entity, count*
Result: *take_from_ent*
begin

    IR_Entity take_from_ent;
    IR_Entity cur_sub_ent;
    cur_sub_ent = target_entity → first();
    while *take_from_ent == ∅ and cur_sub_ent ≠ ∅* **do**
        take_from_ent = MoveUnit (*unit, source_entity, cur_sub_ent, count+1*);
        cur_sub_ent = cur_sub_ent → next();

    if *count == 0 and take_from_ent == ∅* **then**
        take_from_ent = MoveUnit (*unit, source_entity, target_entity* → getParent(),0);

    if *take_from_ent* **then**
        MoveUnitOneLevel (*unit, take_from_ent, target_entity*);
        take_from_ent = target_entity;

    if *source_entity == target_entity* **then** *take_from_ent = target_entity;*
    return *take_from_ent;*

end

---

path-endpoint and path-signal (while traversing away from the IR-unit) are determined. In case the IR-unit is moving up in the hierarchy, these path-port and path-signals may correspond to the segment of IR-path internal to the source entity. In that case, the path segment is removed. If it is not a path-port, indicating it is connected to another IR-unit inside the source entity via a path-endpoint, a new path segment is inserted. If the IR-unit is moving down in the hierarchy, similarly, the next port is checked to be part of the target entity or not. If it belongs to the target entity, a path segment becomes redundant. That is removed. Alternatively, a new path segment crossing the boundaries of the target entity is created.

The algorithm 6.2 is called from the scope of the IR-entity. The value of count is initialized to 0. The other parameters of the function *MoveUnit* are the absolute source and target IR-entities for the IR-unit to be moved. The function recursively searches through the complete IR-entity hierarchy for the path to connect between the source and target IR-entities. This path is then traced back by a single-level hierarchical unit movement. The path tracing is essentially performed by moving up and down the IR-entity tree in the first few lines of the algorithm, where all the sub-entities of the current target IR-entity or the parent of the current IR-entity is sought after. By keeping a simple monotonically increasing variable (*count*), it is ensured that the same IR-entity is not visited repeatedly. Once the complete path is traced, the source IR-entity becomes same as the target IR-entity within the scope of current recursion of the algorithm. From that point onwards, the backtracking recursion of the algorithm keeps on calling the one-level unit movement.

6.3.2 Re-Configurable Unit and Pipeline

The partitioning of re-configurable unit and pipeline are nothing but IR-entity movement, which is again based on the unit movement algorithm. The only additional measure to be taken for IR-entity movement is that, before the IR-unit to be moved, the IR-entity needs to be present there. Therefore, a complete hierarchy of the

---

Algorithm 6.3: MoveEntity

Input: *entity_to_move, wrapper_entity*
begin
    | *IR_Entity entity_clone* = cloneEntity(*entity_to_move*);
    | *wrapper_entity* → addEntityToList(*entity_clone*);
    | *IR_Entity sub_ent* = *entity_to_move* → first();
    while *sub_ent* ≠ ∅ **do**
    | | MoveEntity(*sub_ent, entity_clone*);
    | | *sub_ent* = *sub_ent* → next();
    | *IR_Unit sub_comp* = *entity_to_move* → first();
    while *sub_comp* ≠ ∅ **do**
    | | MoveUnit(*sub_comp, entity_to_move, entity_clone*, 0);
    | | *sub_comp* = *sub_comp* → next();
    | *entity_to_move* → getParent() → removeEntityFromList(*entity_to_move*);
end

IR-entity (without any IR-units) is to be first created. Following that, the IR-units within the source IR-entities are moved to the target IR-entities. This is presented as a pseudo-code in the algorithm 6.3. Also to be noted is that while moving the entity out, a wrapper entity is generated. The top-level function is called with the wrapper entity posited outside the base processor.

6.3.3 Register Localization

The localization of a register has the far-reaching benefit of reducing routing cost as well as critical path. Though most commercially available gate-level synthesis tools perform this task in the flattening phase, it certainly requires to be done beforehand for rASIP. This is due to the separated synthesis flow of re-configurable part and base processor part in an rASIP. The FPGA partition of rASIP is processed from RTL level onwards with the FPGA synthesis technology, whereas for the base processor ASIC synthesis is done.

The registers are represented in the IR as IR-units within the complete register file modelled as IR-entity. The entity-level movement is, thus, sufficient to relocate any particular register to the chosen partition. Before performing the movement, it is necessary to ascertain which registers can be moved to the chosen re-configurable partition. This is performed by checking the read-write access of each register. If all the read-write accesses of a particular register originate from the re-configurable block, it is moved there. The same principle applies for the pipeline registers as well. However for the pipeline registers, presence of any control (stall, flush) from the base pipeline controller is checked for absence before the movement.

6.3.4 Decoder Localization

In the IR, the decoders are modelled as IR-units. For each pipeline stage, one decoder IR-unit is created. That decoder is responsible for driving the IR-processes of that stage and possibly pass some decoded signals to the consequent pipeline stages. With decoder localization, the IR-processes transferred to the re-configurable part of the rASIP, takes their corresponding decoder to the re-configurable part as well. This bears two advantages. Firstly, the decoding is kicked off in the re-configurable part by receiving the pipeline register carrying instruction word itself. This allows an easier synchronization between two partitions. This also may reduce the number of interfacing bits if the number of decoded signals is more than the size of the instruction word. Secondly, the part of the decoder being in the re-configurable partition allows the designer much freedom to re-organize and enhance the instruction encoding belonging to that partition. This decoder localization is shown pictorially in the following Fig. 6.6.

Moving the decoder to the re-configurable partition does not work straightforwardly by moving IR-units. This is due to the designing principle of having one decoder IR-unit per pipeline stage. This does not pose any problem when an entire pipeline is moved. Nevertheless, this is an issue to solve if an UNIT is dubbed re-configurable as shown in Fig. 6.6. In this case, only a part of the decoder i.e. a part of the IR-unit is to be moved. This problem is solved in two phases. Initially, during the construction of IR itself, a local decoder is created within the re-configurable unit IR-entity. Then the IR-unit is moved while moving the IR-entity. Clearly, the second part of the problem follows the algorithms discussed so far. The first part, on the other hand, requires a priori understanding of the construction of the decoder IR-unit. This is explained in the algorithm 6.4. The pre-existing algorithm of one decoder per one pipeline stage is modified, by which any IR-entity is taken as the input. The IR-entity, in turn, acts as a host for the local decoder.

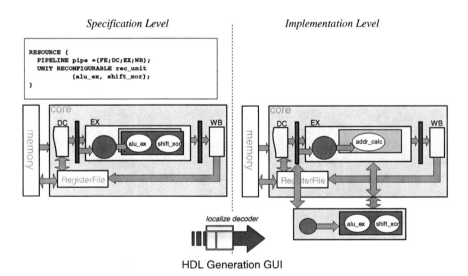

Fig. 6.6 Decoder localization in LISA

Algorithm 6.4: BuildDecoder

Input: *rec_unit_entity*
begin

    I R_Unit decoder = new *I R_Unit*;
    rec_unit_entity → addUnitToList(*decoder*);
    bool *write_to_path*;

    foreach *lisa_operation* ∈ *rec_unit_entity* **do**
        write_to_path = true;
        I R_Path activation = new *I R_Path*(*lisa_operation*);
        decoder → addPathToList(*activation, write_to_path*);
        if *lisa_operation* rightarrow isInCodingTree() **then**
            I R_Path coding = new *I R_Path*(*lisa_operation*);
            decoder → addPathToList(*coding, write_to_path*);
        Operation parent_op = *lisa_operation* → getParent();
        while *parent_op* ≠ ∅ **do**
            if *parent_op* → getEntity() ≠ *rec_unit_entity* **then**
                write_to_path = false;
                I R_Path parent_activation = getPathFromEntity(*parent_op* → getEntity());
                decoder → addPathToList(*parent_activation, write_to_path*);
                break;
            parent_op = *parent_op* → getParent();

end

The decoder is built up by gathering IR-paths, which it should either write to or read from. The IR-paths of type activation are necessary for driving the IR-processes, which again are formed out of the LISA operations. In the algorithm 6.4, it is shown that the decoder is first instantiated within the given IR-entity. Following that, all the LISA operations belonging to the re-configurable unit are traversed. The activation IR-paths for the operations are appended to the current decoder. If the operation contributes to the coding tree, then it needs to be matched with the coding pattern. The IR-path corresponding to that is appended, too. The LISA operation under current iteration require the activation from its parent operations to be properly decoded. The activation IR-paths for those can be created at the current entity or can be read from entities located earlier in the pipeline data flow. Nevertheless, it requires addition of IR-paths in either read or write mode. The addition of these IR-paths are performed until it requires an incoming IR-path. Obviously it is not possible to have the ancestry of the LISA operations to return to the current entity unless, erroneously, circular LISA operation chain is created. The subsequent phase of formation of IR-processes for the decoder IR-unit is straightforward. That is performed by taking each appended IR-path (coding), creating the if-else block out of it by comparing with the instruction register in the pipeline data-flow, and assigning high signal to it. For IR-path activation, it is either read from another entity or is driven to high value by checking the IR-path coding signals.

6.4 Latency : Multiple Clock Domain Implementation

As investigated in a recent work [158], the clock domain allocation of various system components in a SoC is an important design decision leading to various power, performance alternatives. rASIPs do offer the possibility to have different clock domain allocations, too, although in a small scale. For fine-grained general-purpose

FPGA architectures, it is typical to have the same datapath synthesized at a lower clock frequency compared to the ASIC synthesis. Designer can run the entire rASIP at a lower clock frequency as the FPGA commands or can run the rASIP with two different, synchronized clocks. These options lead to different performance results as reported in [159]. To experiment with these choices, the latency keyword is introduced in LISA 3.0 and a corresponding enhancement in the instruction-set simulator is presented. The effect of latency can be captured using the following RTL implementation.

6.4.1 Synchronizing Clock Domains Using Flancter

To ensure the correct performance of a flip-flop, the *setup* and *hold* time restrictions of it must be followed by the incoming signal. While setup time requires the incoming signal to be stable for a while *before* the clocking edge, the hold time requires the signal to be stable *after* it. Without these restrictions, the output of the flip-flop reaches a state of meta-stability and possibly to an erroneous value. Ensuring the timing restrictions is particularly problematic for multiple clock domain implementations. A two-stage synchronizing circuit between two clock domains is proposed at [160]. However, this circuit is not practical for transferring large number of signals as noted in [161]. For single-bit transition, the 2-stage flip-flop-based synchronizing circuit can take data safely across clock domains. For multi-bit transition, the individual bits may become stable at various instances, making the synchronizing circuit much less effective. A solution based on handshake mechanism, called *flancter*, for transferring data across clock domains is proposed in [162]. In the current context, flancter serves the purpose best due to its generality and robustness. The working principle of flancter is explained in the following.

Figure 6.7 shows how the hand-shake mechanism can be employed to transfer the data between two clock domains. The flag represents the current state of the data transfer. When a new data is available, it is set by the source clock domain. When the data is read, it is reset by the target clock domain. By having the access to the current state, a new value is written or read by the source or target clock domains respectively. The important part of the design is to create a circuit for setting, resetting and reading the flag - operable from both the clock domains. A corresponding circuit, proposed in [162], is shown in the following Fig. 6.8. The flancter circuit

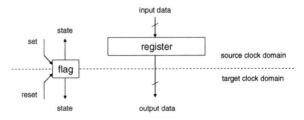

Fig. 6.7 Handshake mechanism of data-transfer

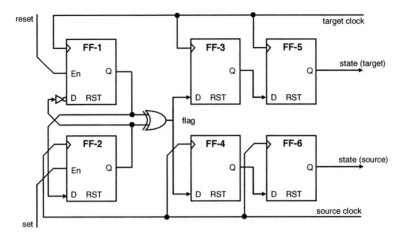

Fig. 6.8 Flancter circuit

itself consists of two flip-flops with enable (FF-1 and FF-2), an inverter and an xor gate. The enable signals are coming from different clock domains to set or reset the flancter circuit's output flag, as shown in the previous Fig. 6.7. The output flag is read individually in two different clock domain after synchronizing it via two consecutive flip-flops (FF-3, FF-5 and FF-4, FF-6). The entire handshake mechanism with flancter ensures reliable data-transfer across two asynchronous clock domains.

6.4.2 Modifications for rASIP

In case of rASIPs, a narrower case is dealt with, where the two clock domains are synchronized by having a clock divider in place. Under such a scenario, the flancter circuit can be avoided for transferring data from the slower clock domain to the faster clock domain (assuming the clocks are generated by the same source and negligible clock skew is present). For transferring the data from faster to slower clock domain, a buffer is set up in the faster clock domain, which holds the data as long as the slower clock domain does not reset the flag via flancter circuit.

   Note that, this implementation can be used only when the re-configurable block's RTL description is simulated with the base processor's RTL description. In case of the re-configurable block being synthesized on the coarse-grained FPGA to generate configuration bitstream, the latency is not explicitly considered.

6.5 Automatic Interface Generation and Storage

In order to ensure that the interface created between the base processor and the re-configurable block during pre-fabrication implementation is maintained during post-fabrication implementation, it is imperative to store the interface. In this work,

Table 6.2 Attributes of IR-paths stored in the XML format

IR-path	Common attributes	Special attributes
register_resource	*Data Type, Direction, Signal Name*	*Resource Name, Read_Or_Written*
memory_resource		*Resource Name, Read_Or_Written*
signal_resource		*Resource Name, Read_Or_Written*
coding		*Operation_Chain*
activation		*Operation_Name*
pipe_reg_control_stall		
pipe_reg_control_flush		
reset		
clock		

the interface is generated directly during IR generation in form of IR-paths. As observed in previously detailed algorithms, the IR-unit as well as the IR-entity is equipped with the list of IR-paths connected to it. The automatic interface generation algorithm looks through the IR-entity hierarchy starting from the top-level and finds out the re-configurable IR-entity by a special tag attached to it during the IR construction process. The list of IR-paths for this particular IR-entity is then traversed and stored.

The storage of the interface needs to be done in a format, which can be later read during post-fabrication exploration of rASIP. In this case, it is chosen to be of XML format. For each kind of IR-path, several attributes of it are stored, which uniquely represents the IR-path. The following Table 6.2 summarizes the attributes stored for different IR-paths. The common attributes are listed in the second column, whereas the special attributes are listed in the third column.

All the IR-paths have data-type, direction (towards the re-configurable block or away from it) and signal name as common attributes. However, these attributes do not sufficiently represent the uniqueness of IR-paths like register, memory and signal resources. For those, the name of the resource as well as is mode of access (read or written) is stored in the XML description of interface. The IR-path coding and IR-path activation are both part of the IR-unit decoder. The IR-path coding is used for matching the op-codes whereas, on the basis of these IR-path codings one IR-path activation is derived for one LISA operation. Every IR-path activation is, thus, attached to one LISA operation. The coding IR-path, on the other hand, can result in different outputs when put into different contexts. Therefore, it is useful to retain the complete operation context (or operation chain) under which the current IR-path coding is accessed.

6.6 Background for FPGA Implementation

As the FPGA is an integrated part of a rASIP architecture, it is extremely important to be able to design and/or choose the specific FPGA architecture tailored for the target application(s). In the proposed pre-fabrication flow an FPGA exploration phase is, thus, incorporated. Two things are important for performing the FPGA exploration at this phase. Firstly, the identification of portions of the architecture to be mapped onto the FPGA. This is covered under the partitioning features of LISA language extension. A Data Flow Graph (DFG) representation of the specified partition is automatically constructed. This acts as an input to the FPGA-based mapping tools. Secondly, it is crucial to model the structural topology, elements and network connections of the target FPGA. This is also covered by the LISA language extensions to capture the FPGA description. Following the elaboration of the partitioning methods and FPGA description formats, it is the time for having the partition mapped, placed and routed on the target FPGA. The mapping algorithm, used in this flow, is derived from the existing state-of-the-art delay-optimal mapping algorithm from the domain of fine-grained FPGA architectures [163]. The algorithm is adapted for a generic input FPGA description, both coarse-grained and fine-grained. The mapping results must be presented in an architecture-independent fashion for navigating the entire design space exploration process. Similarly a generic placement and routing algorithm, also derived from the domain of fine-grained FPGA, is presented. In the following subsections, these steps are elaborated in detail.

6.6.1 Re-Configurable Part of rASIP : DFG-Based Representation

Here a graph-based data-structure is presented, which captures the partition – to be mapped on FPGA – in an intermediate level and is provided as an input to the FPGA mapping tool. Formally, the graph vertices of $G_{DFG} = \langle V_{op}, E_{ic} \rangle$ are the basic *operators* for data manipulation e.g. additions while edges represent the flow of unchanged data in form of *interconnections* of inputs and outputs.

Operators : The following list summarizes the basic classes of operators represented by graph vertices. This special choice of vertices allows us to represent the data flow information in a level between RTL and logic-level representation. Note that the unary operators are considered as a special case of non-commutative n-ary operator class.

- Commutative n-ary Operator
- Noncommutative n-ary Operator
- Read Access to Registers and Memories
- Write Access to Registers and Memories
- Read and Write Access to Array of Variable
- Multiplexer

Interconnections : Interconnections represent the data flow on symbol-level, which again can be suitably decomposed into bit-level representation. The information about the data type transferred is given by an annotation to the interconnection. Bit range subscriptions are included into the interconnection information, too. A direct benefit of this approach is the possibility to encapsulate shift operators with a constant shift amount in bit ranges, thereby reducing the graph complexity.

DFG Creation : For the data-path as well as the control-path the initial internal representation, after parsing the LISA description, is with a Control Data Flow Graph (CDFG). For the purpose of the FPGA library mapping and other optimizations, this is translated to the aforementioned DFG representation. There are three basic differences between CDFGs and pure data flow graphs creating the challenge for the translation algorithm:

- *Control Statements* in CDFGs are block orientated, whereas in DFGs multiplexers are instantiated for each signal and variable written in a CDFG block.
- *Variable Dependencies* given in the sequential statements of CDFGs needs to be replaced by direct interconnects between operator inputs and outputs.
- *Assignments to RTL signals* may occur several times in CDFGs, but in DFGs they have to be concentrated by the use of multiplexers into one physical write access.

To meet these challenges, the translation needs to imitate the sequential execution of a CDFG by tracking the current assignments to variables and signals. Assignments will be represented by the association of the variable or signal with interconnects originating from the output of data flow graph vertices that compute the written value.

Translation of Contexts : The current values for variable and signal assignments are stored in translation contexts. Such a context will be created locally each time a conditional block is entered, using the assignments from the parent context as initialization. Contexts are used for conditional blocks within a conditional statement. The conditional statement checks the conditional expression for different cases, 0 and 1 for if-statements and arbitrary values for switch-case statements. One or more condition values are assigned to each block and therefore to the translation context of each block. The condition values assigned to translation blocks are used when the contexts created for the different blocks of a conditional statement are merged into the parent context. At this time multiplexers are inserted for each variable modified within any block using the condition as control signal and the condition values assigned to the different contexts as the multiplexer input selection criteria.

An example for the context usage is given in the Fig. 6.9. When the analysis reaches the switch statement, its parent context is given by X_p using the output connections $C_{out,add}$ and $C_{out,mul}$ as current values for the variables a and b respectively. The analysis of the switch statement creates the contexts $X_{sw,i}$ for each case statement, the cases 1 and 2 are combined into a single context. The default context is given by the parent context implicitly. From the switch statement, two individual multiplexers are generated for the variables a and b, providing their new input connections $C_{out,mux,a}$ and $C_{out,mux,b}$ for the updated parent context X_p'.

Multiplexer Creation : A multiplexer is created whenever the translation contexts of a conditional statement are combined into their parent context. For this purpose the set V of all variables written within any context is collected. For each variable v, the set CC_v of all pairs (cc, C_{cc}) of condition values cc and associated input connections C_{cc} is created, including the default assignment coming from the

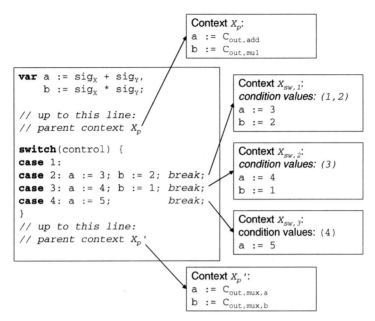

Fig. 6.9 Example of context translation

previous input connection stored in the parent context X_p. If there is no assignment in any parent context, the initialization with zero is assumed as default input. This forced initialization is necessary to avoid undefined signals. In the next step, for each variable $v \in V$ the individual multiplexer can be generated from CC_v, its output connection $C_{out,mux,v}$ is inserted as current value of v into the translation context X_p. The result of the multiplexer generation from the example given in Fig. 6.9 is depicted in Fig. 6.10. The assignments to a and b from the parent context X_p are used as default input for both multiplexers.

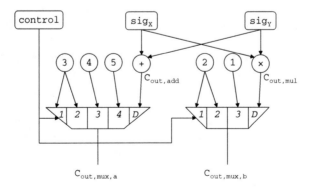

Fig. 6.10 Multiplexer generation from translation of contexts

6.6.2 LISA Language Extensions for FPGA Description : Recapitulation

LISA language extensions for modeling the FPGA are elaborately presented in the Chapter 4. Here, it is briefly outlined to maintain the focus of current discussion. The proposed FPGA description format is composed of three sections, as outlined in the following.

Element A logic block in LISA FPGA description can be described using the ELEMENT keyword. Within an element, the I/O ports, the behavior of the logic block and temporary storages are defined. The behavior section can specify a flexible list of operators via the keyword OPERATOR_LIST. The I/O ports can also have considerable flexibility by having an attribute. The attribute enables an I/O port to be registered or to be directly connected to the external system.

Topology The TOPOLOGY section of LISA FPGA description is used for defining the structural topology of the FPGA. Within the topology section, one-dimensional or two-dimensional arrangements of the logic blocks can be made using special keywords.

Connectivity The interconnects between the logic blocks of the FPGA and between different hierarchies of the FPGA structure can be defined within this section. The connectivity is grossly guided by one more specific connectivity styles (e.g. mesh, nearest neighbour) and fine-tuned by specific connection rules.

6.6.3 Internal Storage of the FPGA Structural Description

To seamlessly access the structural information about the FPGA block, an internal data-structure is constructed. The data-structure is organized in tune with the above-mentioned three sections of the FPGA description. In synchronization with the general IR, the FPGA internal representation is noted as FIR i.e. FPGA Internal Representation.

Elements : The *elements*, which represent the fundamental blocks of the FPGA operation, are captured in form of pattern graphs to be used for library mapping. A pattern graph can be formally represented as a directed acyclic data-flow graph $G_{PAT} = \langle V_{op}, E_{ic} \rangle$, where the basic operators are captured as nodes and the data dependencies are represented as interconnect edges. As outlined in detail in the previous chapter, the elements contain ports (which are connected hierarchically to clusters) and behavior written in C. As mentioned, the ports can be attributed to *register* or *bypass* those, which eventually adds up to the behavior. The element's behavior is parsed and represented internally in form of a DFG. The DFG representation of an element is kept in entirety for RTL synthesis. For performing mapping, this is again decomposed into smaller pattern graphs. The DFG representation of

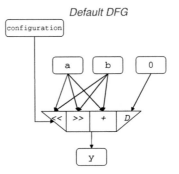

FPGA element description

```
ELEMENT alu {
  PORT{
    IN unsigned<16> a,b;
    OUT unsigned<16> y;
  }
  ATTRIBUTES {
    REGISTER(y);
    BYPASS(y);
  }
  BEHAVIOR {
    OPERATOR_LIST op = {<<,>>,+};
    y = a op b;
  }
}
```

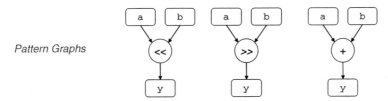

Pattern Graphs

Fig. 6.11 Pattern graph generation from elements

the element's behavior as well as those behaviors resulting from the ports' attributes are stored in FIR-Process much similar to the general IR-Process for the base processor's internal representation. These pattern graphs together constitute the pattern library to be fed as an input during the FPGA library mapping. The formation of pattern graph from a given element description is shown in the Fig. 6.11.

Note that, a single element can give rise to several pattern graphs as shown. The parsing of C code of the elements' behavior section and their consequent DFG formation works in the same way as elaborated before for the DFG generation of the re-configurable part of the LISA description. Actually, the same data-structure is used for storing the pattern graphs as well as the input DFG for re-configurable part of the processor. Once the complete DFG for the element is formed, it is decomposed to construct several small patterns. This process is guided by the pseudo-code furnished at algorithm 6.5. Currently, the pattern generation process assumes that at most one OPERATOR_LIST is allowed inside the behavior of an FPGA element. That is reflected in the algorithm, too. At the first phase of the algorithm, the number of patterns are evaluated by looking for multiplexers, which are created from OPERATOR_LISTs. During the second phase, for the multiplexer, one by one case is dealt with. For the chosen case, the source node and the target nodes are connected together. Following that, the newly built graph is cleared up by finding what is really connected to the primary outputs. After several iterations, one for each case of the multiplexer, a list of pattern graphs is prepared.

Topology : The *topology* is captured in form of a directed n-ary tree $T_{topo} = \langle E, R \rangle$. Here, the vertex set $E = (E_1, E_2, .., E_n)$ represents the entities/modules

---

Algorithm 6.5: Generate_Pattern_Graphs

Input: $DFG\_element$
Result: $DFG\_Pattern\_List$
begin

  foreach $node \in DFG\_element$ **do**
    if $node \to$ getType() == $OPERATOR\_LIST\_MUX$ **then**
      $pattern\_count = node \to$ getCasesCount();

  for $index = 1$ **to** $pattern\_count$ **do**
    G_{PAT} $DFG\_Pattern$;
    foreach $node \in DFG\_element$ **do**
      if $node \to$ getType() $\neq OPERATOR\_LIST\_MUX$ **then**
        $new\_node = node \to$ cloneNodeEdge();
        $DFG\_Pattern$ append(new_node);

    foreach $node \in DFG\_element$ **do**
      if $node \to$ getType() == $OPERATOR\_LIST\_MUX$ **then**
        $source\_node = node \to$ getCase($index$) $\to$ getSource();
        $target\_node = node \to$ getTarget();
        $new\_target\_node = DFG\_Pattern \to$ findCorrespondingNode($target\_node$);
        $new\_source\_node = DFG\_Pattern \to$ findCorrespondingNode($source\_node$);
        $new\_source\_node \to$ addOutput(new_target_node);

    foreach $PO\_node \in DFG\_element$ **do**
      $new\_PO\_node = DFG\_Pattern \to$ findCorrespondingNode(PO_node);
      $DFG\_Pattern \to$ backtrackAndMark(new_PO_node);
    $DFG\_Pattern \to$ clearUnmarkedNodeEdge();
    $DFG\_Pattern\_List$ append($DFG\_Pattern$);

  return $DFG\_Pattern\_List$;

end

(FIR-Entity) in the FPGA. The parent-child relationship, which forms the edge set R, is captured by the directed edge (E_i, E_k) between two entities. Furthermore, each entity E_i stores the port information (e.g. direction, width, sign) in form of a list $P_e = (P_1, P_2, .., P_m)$. Each element is entitled to an entity of its own. Therefore, an FIR-Entity is constituted by four members as noted in the following.

1. Set of children FIR-Entities E_e, which empty in case of an element.
2. List of FIR-Ports P_e.
3. Set of pattern graphs G_{PAT_e}, which is empty in case of the entity being non-elementary.
4. Set of FIR-Processes $Proc_e$, representing the behavior of element, behavior of attributed ports etc.

Connectivity : The *connectivity* is stored in a hash-table data-structure, where the entity/module name $Name_e$ acts as the key to the hash-table. The connectivity information $Conn_e$, within the scope of the given FIR-Entity, is stored as the data of the hash-table with the name of FIR-Entity acting as the key. $Conn_e$ contains firstly, the connectivity style and stride for the entity e and secondly, the list of basic rules $Rule_e$ specifying the connection restrictions within the set of entities (e, E_e). One specific rule $Rule_{e_i}$ within $Rule_e$ deals about the port connectivity constraints between a particular source e_{isrc} and destination e_{idst} entities. Every rule contains several port to port connection possibilities, which the FPGA Synthesis tool (as well as the FPGA RTL Implementation tool) must adhere by. Apart from these set of rules, the connectivity style (mesh, nearest neighbour) and the connectivity stride (1-hop, 2-hop) are also stored in the context of one FIR-Entity.

6.6.4 Coarse-Grained FPGA Architecture Implementation from High-Level Specification : Related Work

The modelling of a flexible hardware system by its nature incurs strong performance loss. To minimize this loss as much as possible, FPGAs are mostly designed at low abstraction level with gate-level, transistor-level or physical-level optimizations. This holds equally true for both fine-grained and coarse-grained FPGAs. These optimizations are generally not possible to trigger via commercial high-level synthesis tools thus, leaving FPGA specification at high-level serving the sole purpose of enabling the mapping algorithms. Recently, an RTL abstraction of FPGA is proposed [164] for simulation purpose only. Moreover, this does not provide any way to derive this RTL abstraction automatically. There are some attempts to synthesize FPGA structures directly from high-level specification to layout level using custom data path generator [165]. Though the effort is commendable, it offers increased complications because of debugging the implementation at low level of abstraction. The issue with selection of functional units for coarse-grained FPGA development has been approached at [142] and [166]. However, the path to implementation is not outlined there.

6.7 FPGA Implementation : RTL Synthesis

RTL synthesis from the high-level FPGA structural description goes through the initial phase of populating the FPGA-IR elements, followed by the back-end to map it to HDL description. This process is explained in the following using pseudo code.

The above pseudo code from function 6.6 summarizes the overall process of FPGA RTL synthesis. At the beginning, the FPGA description from LISA is parsed. Immediately followed by the parsing, the FIR-Entity hierarchy is established, thus returning the top-level entity. The next function of importance is the generation of DFG for the behavior code inside the *elements*. The DFG representation is obtained by converting the CDFG one, which is the default outcome of the parsing process. This conversion from CDFG to DFG is done as explained previously. At this point, the FPGA-IR is completely set up with entity hierarchy as well as internal DFG representation. Then, the connectivity section from the FPGA Description is accessed to establish the link between the ports of FIR-Entities. In the following pseudo code, this function is elaborated.

As outlined in the overall algorithm 6.6, the function connectPaths is called for the first time from the top-level FIR-Entity. Within this function, it is recursively called for each of its sub-entities. For the current entity, the connectivity style $Conn_e$ is obtained from the hash-table. Within one single connectivity, a large

---

Algorithm 6.6: FPGA_RTL_Synthesis

Input: *FPGA_Description*
Result: *HDL_Description*
begin

    FIR-Port_List *top_conf_port_list*;
    FIR-Entity E_top = parseFPGADescription(*FPGA_Description*);
    generateDFG(E_top);
    connectPaths(E_top);
    connectConfiguration(E_top, *top_conf_port_list*);
    $HDL\_Description$ = generateHDL(E_top);

end

---

number of basic rules ($Rule_e$) exist. The main part of algorithm 6.7 dwells over
each such rule. For every rule, the source and destination FIR-Entity as well as the
port names are obtained. Note that, one rule can connect one single source port
to a list of destination ports along with a choice of a specific range or subscription.
These details are omitted here for the sake of simplicity. Nevertheless, the algorithm
principle is still very much the same as depicted here. The source and destination
FIR-Entity of a single rule can be a child sub-entity of the current FIR-Entity or the
current FIR-Entity itself. Moreover, the source and destination FIR-Entities cannot
be the same instance. Once the source and the destination FIR-Entities are established, it is checked whether the destination FIR-Entity satisfies the connectivity
style (e.g. point-to-point, mesh) and the connectivity stride (e.g. single stride, dou-
ble stride) with respect to the source FIR-Entity. This check is done via the function call allowedByStyle, which verifies their relative position and distance to
adhere by the connectivity specifications. On the satisfaction of these constraints,

---

Algorithm 6.7: connectPaths

Input: *curr_entity*
begin

    foreach *child_entity* ∈ *curr_entity* **do**
        connectPaths(*child_entity*);

    $Conn_e$ *curr_conn* = getConnectivity(*curr_entity* → getName());
    foreach *rule_i* ∈ *curr_conn* **do**
        rule_i → getSrcName(*src_ent_name*, *src_port_name*);
        rule_i → getDstName(*dst_ent_name*, *dst_port_name*);
        if *curr_entity* → getName() == *src_ent_name* **then**
            foreach *dst_entity* ∈ *curr_entity* and *dst_entity* → getName() == *dst_ent_name* **do**
                if *dst_entity* → allowedByStyle(*curr_entity*, *curr_conn*) **then** connectPort(*curr_entity*,
                src_port_name, *dst_entity*, *dst_port_name*);

        else
            foreach *src_entity* ∈ *curr_entity* and *src_entity* → getName() == *src_ent_name* **do**
                if *curr_entity* → getName() == *dst_ent_name* and
                curr_entity → allowedByStyle(*src_entity*, *curr_conn*) **then**
                connectPort(*curr_entity*, *src_port_name*, *dst_entity*, *dst_port_name*);
            else
                foreach *dst_entity* ∈ *curr_entity* and *dst_entity* ≠ *src_entity* **do**
                    if *dst_entity* → getName() == *dst_ent_name* and
                    dst_entity → allowedByStyle(*src_entity*, *curr_conn*) **then**
                    connectPort(*curr_entity*, *src_port_name*, *dst_entity*, *dst_port_name*);

end

---

the FIR-Ports of these FIR-Entities are connected. For connecting these FIR-Ports, internally, data-structures similar to IR-paths are used.

As a major extension to the previous IR-path concept for connecting FIR-Ports, a special data-structure element named FIR-Path_Hub is conceived. Two FIR-Ports are connected during the `connectPort` function call. For each FIR-Port distributing to more than one FIR-Port as well as for each FIR-Port collecting from more than one FIR-Port, an FIR-Path_Hub (either of distributor or of collector type respectively) is constructed. Two FIR-Path_Hubs are connected via an FIR-Path_Wire. During the drive of back-end to map these representations to HDL, the FIR-Path_Hub of collector type is converted to a RTL description of many-to-one multiplexer. These multiplexer elements are exactly those, which are responsible for providing the flexibility to the routing network of the FPGA. Note that, in the physical implementation, the multiplexers are modeled as pass-transistors to optimize area, delay and power. These multiplexers or FIR-Path_Hubs (as those are represented before back-end) are controlled by configuration bits coming from the top-level FIR-Entity.

The following pseudo code (algorithm 6.8) of the function `connect Configuration` dwells upon the connecting of configuration bits from the top-level entity to each FIR-Path_Hub as well as other elements which present an ability to be re-configured post fabrication.

Similar to the previous `connectPaths` function, `connectConfiguration` also begins with the top-level FIR-Entity as the current FIR-Entity. For each entity, configuration FIR-Ports or FIR-Paths are sought recursively in children FIR-Entities or FIR-Processes respectively. The configuration FIR-Paths in the FIR-Processes are initially built up either during `parseFPGADescription` function (for OPERATOR_LIST or for attributed FIR-Ports) or during the `connectPaths` function (for FIR-Path_Hub of collector type). The configuration FIR-ports and FIR-paths from children components are assembled to prepare a single FIR-Port of configuration

---

Algorithm 6.8: connectConfiguration

Input: *curr_entity*, *curr_conf_port_list*
Result: *curr_conf_port_list*
begin
    unsigned int *total_conf_width* = 0;

    foreach *child_entity* ∈ *curr_entity* **do**   connectConfiguration(*child_entity*, *ent_conf_port_list*);
    foreach *child_process* ∈ *curr_entity* **do**   getProcessConfiguration(*child_process*,
    proc_conf_path_list);
    foreach *child_port* ∈ *ent_conf_port_list* **do**   *total_conf_width* += *child_port* → getWidth();
    foreach *child_path* ∈ *proc_conf_path_list* **do**   *total_conf_width* += *child_path* → getWidth();
    if *total_conf_width* ≠ 0 **then**
        Char∗ *curr_conf_name* = "conf_" + *curr_entity* → getName();
        curr_conf_port = *curr_entity* → appendConfigurationPort(*curr_conf_name*, *total_conf_width*);
        src_upper = *total_conf_width* - 1;
        foreach *child_port* ∈ *ent_conf_port_list* **do**
            │ *src_lower* = *src_upper* - *child_port* → getWidth() + 1;
            └ connectConfPort(*curr_conf_port*, *src_upper*, *src_lower*, *child_port*);

        foreach *child_path* ∈ *proc_conf_path_list* **do**
            │ *src_lower* = *src_upper* - *child_path* → getWidth() + 1;
            └ connectConfPath(*curr_conf_port*, *src_upper*, *src_lower*, *child_path*);

        curr_conf_port_list → appendToList(*curr_conf_port*);
    return *curr_conf_port_list*;
end

type for the current FIR-Entity. Thereafter, the total width of the current FIR-Port is distributed among the children components' FIR-Ports and/or FIR-Paths.

The back-end function `generateHDL` works in the similar fashion as for the basic LISA-based HDL generation. It performs a language-dependent type propagation before writing out the HDL description. A key difference between the LISA processor description and the FPGA description is that for the latter, same entities are instantiated multiple times to build a regular structure. The back-end, therefore, visits only one instance of every FIR-Entity.

6.7.1 Synthesis Beyond RTL

In this book, a new FPGA modelling abstraction is proposed for the purpose of design space exploration of coarse-grained FPGA blocks. As a result of the FPGA implementation, the RTL description of the FPGA is obtained. The question, however, remains that how to implement the FPGA RTL description physically. This problem is not addressed within the scope of this book. Rather, an overview of the problem along with possible future road map is presented.

The most area-consuming component of modern FPGAs are the interconnects. Huge interconnect network is what allows the tremendous flexibility of FPGAs. As a cost of that, higher area, timing and power is often consumed. Coarse-grained FPGA structures tries to reduce the interconnect structure (making it more application-specific), thus improving performance [91]. Nevertheless, the interconnect architecture maintains its dominance. Obviously, while modelling the interconnect architecture physically, absolute care is taken to curb redundancy. Apart from regular wiring and switch-box structures, multiplexers take up the major portion of the interconnect architecture of the FPGA. A typical transistor-level multiplexer implementation (source [164]) is shown in the left side of the above Fig. 6.12. The complete implementation of a 2-level multiplexer is done using pass transistors. The gray boxes controlling the pass transistors are controlling inputs, which allow different input data to be available at the output. To model the same implementation

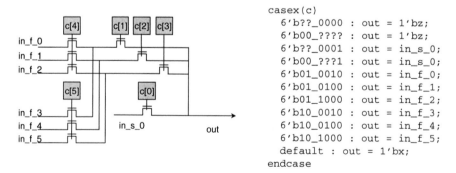

Fig. 6.12 Proposed RTL abstraction for FPGA routing logic (redrawn from source [164])

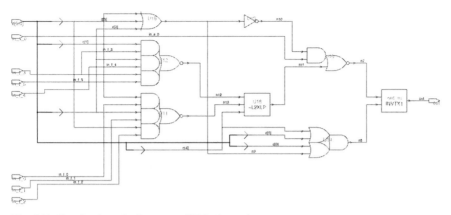

Fig. 6.13 Gate-level synthesis output of RTL abstraction

in RTL abstraction, a form has been proposed also in [164]. This takes into account the possible contention and high impedance states of the output.

Once gate-level synthesis of the RTL description presented in Fig. 6.12 is performed, the resulting logic-level structure, as shown in Fig. 6.13, is obtained. Clearly, this is not anywhere near to what from the abstraction originated. The gate-level synthesis is performed using latest commercial tools [139] with 90 nm technology library. The failure to realize the original multiplexer structure can be attributed to firstly, the inability of gate-level synthesis tools to see beyond a gate (e.g. failure to instantiate pass transistors) and secondly, the un-availability of special elements in the technology library to model such a multiplexing behavior optimally. Both of these problems can be addressed in the future research of FPGA implementation. While high-level modelling of FPGA eases design space exploration, physical implementation is sought after for high performance of the newly designed FPGA. This gap can be closed by either developing sophisticated gate-level synthesis tools or by maintaining a separate design abstraction of only simulation purpose. The later one is what proposed at [164]. An interesting approach is to utilize full custom data path generator [165] for generating the FPGA from a high level of abstraction. This is done in [91]. However, this lacks a methodology to map the application automatically to the FPGA. Furthermore, no high level simulation framework means the complete verification is performed only at transistor level, which is tedious.

As of now, the FPGA implementation beyond RTL is done with the commercial gate-level synthesis tools. The resultant figures represent the trend and not the absolute value, would it be implemented physically with optimizations common in FPGA.

6.8 Background for Synthesis on FPGA

A fundamentally different problem compared to what is discussed so far is *Synthesis on FPGA* or also referred to as *FPGA Synthesis*. During FPGA implementation, the key issue is to model the FPGA structure itself in high-level of abstraction with consequent RTL generation of the entire structure. For FPGA synthesis, the question is to map a given control-data-flow description (given in form of a high-level language) on the existing FPGA structure. For this purpose, the existing FPGA elements have to be used. Moreover, the FPGA connectivity constraints as well as the FPGA topology description needs to be respected. This is a widely researched area, which continuously exchange concepts with another similar field i.e. ASIC synthesis. In both the fields of synthesis, a library of basic elements are provided for mapping the control-data-flow description. While for ASIC synthesis the designer can arrange the elements in any order, for FPGA synthesis certain restrictions on ordering the elements as well as the choice of elements are imposed. The most flexible type of basic element offered by an FPGA is Look-Up Tables or LUTs. An n-bit LUT is able to encode any n-input boolean function by modeling such functions as truth tables. Additional flexibility comes from the hugely diverse routing networks. Together, these constitutes the fine-grained FPGAs, so termed because of the fine logic granularity of its basic elements (i.e. LUTs). Since the FPGAs found most usage in prototyping a design before eventually running ASIC synthesis, flexibility remained a major issue in early FPGA research. As a result, significant progress have been made in field of fine-grained FPGA synthesis. Recently, with the usage of flexible yet high-performing coarse-grained FPGAs ramping up, coarse-grained FPGA synthesis is receiving strong research attention, too. Before delving into the generic coarse-grained FPGA synthesis algorithms used in this book, a brief overview of the FPGA synthesis technology is presented in the following sub-sections.

The phases of FPGA synthesis are shown in the Fig. 6.14 with a fine-grained LUT-based FPGA used as the target. In the first phase i.e. *Mapping phase*, the boolean functions are covered by the existing elements of FPGA library. In the following *Clustering phase*, the elements are gathered together in clusters of the

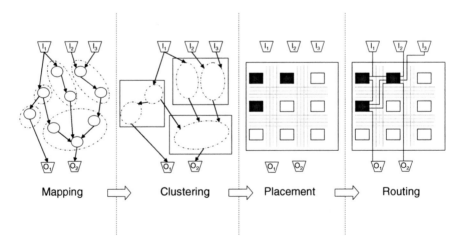

Fig. 6.14 Phases of FPGA synthesis

FPGA structure. In the third phase, which is *Placement phase*, the clusters are placed on the FPGA topology. This is followed by the final *Routing phase*, where the interconnects are joined to the clusters while maintaining functional equivalence with the original boolean network. Numerous algorithms, with optimization focus on area, power or delay have been proposed in the literature. In the following sub-section, the most prominent algorithms an the basic cost models for delay used in those are discussed. For detailed development of FPGA synthesis technology, interested readers may refer to [167].

Delay Model : For organizing the logic blocks of a VLSI circuit hierarchically or among several components, the concept of clustering was first introduced in ASIC synthesis. Clustering is performed as a follow-up of mapping, where the boolean functions are tagged to the elementary library elements. Early mapping algorithms [168] assumed *unit delay model* to perform the clustering. The unit delay model is defined as following.

1. All gate (or basic library element) delays are zero.
2. No delay is encountered on an interconnection linking two gates internal to a cluster.
3. A delay of one time unit is encountered on an interconnection linking two gates in different clusters.

Later, the delay model was improved to *general delay model* [169] with the argument that, the intra-cluster wires and/or the gate delays also can contribute comparably to the inter-cluster wire delay when the cluster size is significantly large. The general delay model is, thus, defined as following.

1. Each element v of the library has a delay of $\delta(v)$.
2. A delay of D_{int} is encountered on an interconnection linking two elements internal to a cluster.
3. A delay of D_{ext} is encountered on an interconnection linking two elements in different clusters.

Fine-grained FPGA Synthesis Algorithms for Mapping : Fine-grained FPGA synthesis phase of mapping (also referred as technology mapping) deals with finding the functionally equivalent LUT-based representation of a given boolean network so as to minimize delay [170], area [171] or power consumption [172]. The first mapping algorithm to present delay-optimal solution under polynomial time is presented in [173]. However, this algorithm used an unit delay model, which is clearly unrealistic.

Fine-grained FPGA Synthesis Algorithms for Clustering : The problem of clustering can be considered as an issue with partitioning the circuit into several components, satisfying the given design constraints. The design constraints may be the capacity of a cluster (for ASIC, FPGA synthesis) and/or connectivity restrictions (for FPGA synthesis).

The clustering problem was originally solved in polynomial time for unit delay model at [168]. Later, a polynomial time solution for general delay model was presented at [169], which is shown to be optimum under specific conditions. A provably

delay-optimal solution under any monotone clustering constraints was proposed at
[174]. Note that, these clustering solutions are equally applicable to the field of
ASIC as well as FPGA synthesis.

In a recent paper [163], the authors noted that due to the two-phased approach
of FPGA synthesis in mapping and clustering, the results produced are clearly sub-
optimal. This is more so since, the mapping algorithms are performed assuming
the unit delay model. The authors then presented a dynamic-programming based
approach to perform Simultaneous Mapping And Clustering (SMAC) [163]. This
is proved to deliver delay-optimal solution for fine-grained FPGAs. In this book,
SMAC algorithm is used as a starting point to derive a delay-optimal solution for
coarse-grained FPGAs.

FPGA Synthesis Algorithms for Placement and Routing : The algorithms for
performing placement and routing are closely bound to each other. The placement
algorithm finds out the exact physical locations of the logic blocks (clusters) on the
pre-fabricated FPGA hardware and the routing algorithm connects the signals between these logic blocks utilizing the FPGA's available routing resources. Given an
FPGA contains a finite number of routing resources with considerable constraints on
the routing architecture, the quality of final outcome depends strongly on the placement algorithm's solution. Therefore, the algorithms developed for placement contain an inner loop to check whether the placement solution is at all routable or not.

Due to the complexity (NP-hard) of FPGA placement problem, efficient heuristic
solutions are proposed throughout. The current state-of-the-art FPGA placement
algorithms [175, 176] are based on the heuristic of simulated annealing [177]. This
heuristic repeatedly exchanges the position of clusters to improve the placement.
At the inner loop of this algorithm, the routability of the currently placed solution
is measured using architecture-specific heuristics [176]. Since an objective of this
work is to build a generic coarse-grained FPGA synthesis flow, the architecture-
independent flow from [175] is chosen as the starting point here. There, for every
placement, the signals are actually routed instead of measuring the routability in an
architecture-specific manner.

Like mapping and clustering, the problem of routing is also encountered in ASIC
synthesis. In IC design, the routing is divided into two phases, namely, *global* and
detailed routing. In global routing phase, the routing area is reserved, whereas in
detailed routing the pins are actually connected over the routing area. Since the
FPGA routing problem is fundamentally different from IC routing owing to its
highly constrained and restricted routability, therefore in FPGA synthesis the routing is performed in an integrated single step. Nevertheless, few basic algorithms
are borrowed from the ASIC synthesis world. Most notable of them is the shortest
path algorithms. Typically, the global routing of ASIC synthesis is solved by well-
known shortest path algorithms (e.g. maze routing). This approach, however, may
result into unroutable nets for FPGA synthesis. The approaches to solve this are
either rip-up and re-route and/or routing in a specific order. For FPGA synthesis,
[178] performs an early slack calculation for every net from source to sink to obtain upper bounds. The routing is done initially for the complete design. Following
that, selected connections are re-routed to improve the delay. This scheme has a

drawback due to the initial routing order as well as the initially computed fixed slack values. The most prominent routing algorithm (known as *Pathfinder*) for fine-grained FPGA synthesis is presented at [179]. In this algorithm, the congestion of routing resources and delay of critical paths are iteratively balanced. As this algorithm provably yielded best quality results for fine-grained FPGA synthesis [180] so far, this is employed as the routing algorithm for several well-known placement algorithms too [176, 175]. In this work, the *Pathfinder* routing algorithm is used, modifying it only to accept the generic FPGA structure definitions.

6.8.1 Synthesis on Coarse-Grained FPGA Architecture : Related Work

Due to the increasing interest in coarse-grained FPGA architectures, several attempts to explore the architecture choices at high-level have been made. Before proceeding into the details of generic FPGA synthesis technology, it is worth noting those. The problem is different for coarse-grained FPGAs from fine-grained FPGAs as it requires more elaborate description for the processing elements and less elaborate description for the interconnect architecture, generally. In the following we concentrate on coarse-grained FPGA architectures. There are few rASIPs, which made significant advancements in coarse-grained FPGA synthesis technology as well. These are elaborated, too.

There have been many significant attempts to generically define the FPGA architecture and perform mapping, clustering, placement and routing of a given datapath onto it. In many of such cases, the FPGA architecture is partially or completely fixed. However, that may not affect the genericity of mapping algorithms in general. A completely generic approach for exploring the functional units of a coarse-grained FPGA with corresponding mapping algorithm is presented at [181]. In this work, mesh-based grid of processing elements is conceived. For different grid configurations (4×4, 8×8), different interconnect topologies and various functional units inside the processing elements mapping is performed. The mapping algorithm is selected from a set of different topology traversal options. Though this provides a rough hint about the selection of a particular coarse-grained re-configurable block, the mapping algorithm is clearly sub-optimal. No attempts are made to perform placement and routing for the same blocks. In another work by the same authors [182], routing topology is explored. A datapath synthesis system for coarse-grained FPGA is presented at [99]. This allows the FPGA users to write the input datapath in a language called ALE-X (which is strongly oriented to C). The FPGA description is strongly embedded in the mapping, placement and routing algorithms. Post routing, a scheduling is performed to optimally sequence the I/O operations in view of limited bus resources. An interesting approach is adopted for mapping applications to coarse-grained rASIP named GARP [68]. By recognizing that both synthesis and compilation are actually solving the same problem [183], in [68] the FPGA synthesis is performed using similar techniques as found in the the domain of high-level compilers. The input data-flow graph is split into trees and then tree covering algorithm is applied for mapping. Also, it is noted that the placement decisions seriously affect the mapping results. Therefore, a dynamic programming-based placement is performed simultaneously with the module mapping. This work is a precursor of [163] in the domain of fine-grained FPGAs. However, a major disadvantage of the synthesis algorithm in [68] is that, it requires splitting of input graphs into trees and then combining those back together. This comes with sub-optimal result in the global context. Unlike [163], the increased area cost coming from replicated modules are not considered in [68] either. A memory aware mapping algorithm for coarse-grained FPGAs is proposed at [184]. Here, the FPGA architecture is fed into the algorithm in form of an undirected graph. The input datapath is modelled in a data-dependence graph form, where each node of the graph is assigned with its priority. The priority is set by its input requirements and dependence on other nodes. The mapping algorithm works on the principle of list scheduling, where the nodes residing on the critical path are mapped first and so on. For each node, a list of mapping-cum-placement choices are first determined. On the basis of available routing resources, the best choice is selected. In the single-chip system Chameleon [81], tiles of coarse-grained FPGA Montium are used. For Montium [185], a set of template operators are first created. Out of these templates the best ones are chosen, a step which is similar to the FPGA synthesis. There, for each node, all the possible template matches are outlined. From these matches, a conflict graph is created where each node reflects a match and there is an edge if two matches have one or more nodes in common. The weight of a conflict graph node is same as the number of input graph nodes covered by that match. From this conflict graph, the disjoint set of nodes corresponding to one template is identified. The templates which allow more weighty nodes to be selected without overlapping are chosen. This automatically leads to the graph mapping decisions, too. This method, though good for area constraints, overlooks the delay element.

6.9 FPGA Synthesis : Mapping and Clustering

Before engaging with the mapping and clustering solution for the coarse-grained FPGA synthesis, the problem is first formulated clearly. This requires few definitions to be introduced. The input boolean network can be represented as a DAG $G_{DAG} = \langle V_{op}, E_{net} \rangle$, where each node represents a basic operator (available in the FPGA library or possible to decompose into that) and each edge represents a net connecting from the output of an operator to the input of another one. A Primary Input (PI) node is a node with no incoming edges and a Primary Output (PO) node is that with no outgoing edges.

The FPGA architecture, as discussed before, is captured internally in the FPGA-IR. In summary, it contains a list of patterns in form of DAG as well (G_{PAT}). The patterns basically represent the FPGA library elements (or parts of that). The *topology* is captured in form of a directed n-ary tree $T_{topo} = \langle E, R \rangle$ with the vertices representing FIR-Entities and the edges denoting the parent-child relationship. The clusters are represented by FIR-Entities, with a specific set of G_{PAT} within those. The set of patterns for a given cluster E_c can be represented as $S(G_c) = (G_1, G_2, .., G_n)$. The connectivity constraints among the FIR-Ports of various FIR-Entities within the scope of a given FIR-Entity $\mathcal{E}$ are stored captured in a hash-table format in $Conn_{\mathcal{E}}$. Regarding the cost model, a general delay model is assumed with each node of G_{PAT} being associated with a specific delay $\delta(\upsilon)$ along with a particular intra-cluster (D_{int}) and inter-cluster (D_{ext}) delay. On the basis of these definitions the mapping and clustering problem for generic (coarse-grained as well as fine-grained) FPGA synthesis can be formulated as following.

Problem Formulation : Given a particular FPGA architecture with its topology definition T_{topo}, connectivity constraint definition $Conn_{\mathcal{E}}$, pattern library definition G_{PAT}, cluster content definition $S(G_c)$ - determine the mapping of a given input boolean network G_{DAG} on the FPGA architecture - to minimize area, delay or power for a given cost model. In this book, the focus is on delay minimization.

Solution Approach : The interesting thing to observe is that, both the early ASIC synthesis clustering algorithm [169] as well as the latest delay-optimal FPGA synthesis mapping [173] used the same phases of algorithm for generating the LUT-based logic network. The first phase is to *label* the nodes of the input boolean network with the best available delay via dynamic programming and the second phase is to realize the mapping/clustering covers. For our case, the solution approach begins from the state-of-the-art SMAC algorithm, where these phases are also existent. In the following example, the algorithmic principle undertaken in this book is explained pictorially with the Fig. 6.15.

The SMAC receives the input boolean network as an input with several constraints and the delay model. The first phase of SMAC, i.e. the labelling phase, begins from the primary input nodes of the input network traversing to the primary outputs in a level-wise manner. At the very beginning possible pattern graph(s) are identified, which can be used to mimic the functionality of current node in the final FPGA. Following that, SMAC labels the node with a possible solution. A possible solution is marked by a particular cluster type and an unique cluster identification. Given that node is part of a specific cluster, the propagation delay from the primary inputs to this node is measured and also stored as a part of the label for this node. This node can be part of several different clusters started by its parent nodes as long as it satisfies the following regulations.

- ability of an existing cluster to accommodate this node.
- ability of the connectivity constraint to accommodate this node within an existing cluster.

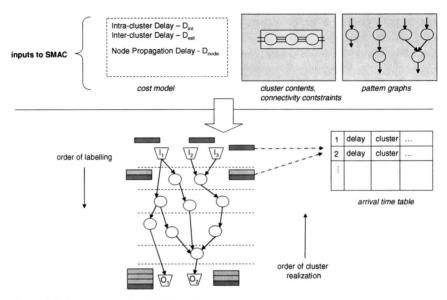

Fig. 6.15 Solution approach of SMAC [163]

In either case, the current node also is used to mark the beginning of a new cluster. The propagation delay in that case is calculated and stored in the label. Therefore, one particular node of the input network can have several labels depending on

1. different possible patterns rooted at this node.
2. different clusters of parent node which are able to accommodate the current node.
3. a new cluster starting from the current node.

While calculating the propagation delay of the current node on the basis of its parent nodes, the maximum propagation delay among the minimum propagation delays of its parents are used as the starting point. Thence, the propagation delay of the current node is calculated according to the following equation (6.1).

$$delay_{node} = \begin{cases} D_{node} + D_{int} + delay_{parent}, & \text{if } node \text{ belongs to } cluster_{parent}. \\ D_{node} + D_{ext} + delay_{parent}, & \text{if } node \text{ marks the beginning of a new} \\ \text{cluster}. \end{cases}$$

(6.1)

As can be observed from the above example, there are definite efforts to retain full genericity in the solution approach. This at the same time covers a wider range than conventional SMAC [163] particularly in maintaining the connectivity constraints and in the graph mapping part. From that perspective, the mapping and clustering algorithm used in this book is referred henceforth as *Coarse-grained Generic SMAC* or *CG-SMAC*.

Labelling : The key data-structure used in the labelling as well as cluster realization phase is termed as *arrival time table*. The arrival time table contains a list of rows, each of which is equipped with a member set. The data-structure is shown in Table 6.3. During labelling, one such arrival time table is maintained for each node of the input DAG.

The first two elements of an Arrival_Time_Row are used to indicate the type of the cluster and the unique identification number of the cluster, to which the current node belongs. The third element, m_propagation_delay is useful to store the delay of propagation for the current node beginning from the primary inputs. The m_bit_matrix_model stores a 2-dimensional matrix structure of binary values, which resembles the physical structure of the current cluster. Each binary value is reflective of an element within the cluster, with the value of 1 indicating that the corresponding element is already occupied. The final element, the m_node_list, contains the nodes of the input G_{DAG}, which are covered by the cluster holding the current node.

The labelling phase of CG-SMAC is explained via the pseudo-codes in algorithm 6.9, algorithm 6.10 and algorithm 6.11. The overall algorithm, presented in algorithm 6.9, is called with all the input constraints due from the FPGA structure and the input boolean network to be mapped. Initially the nodes of the input DAG is sorted level-wise, where from the Primary Input (PI) nodes are obtained. For each of these PI nodes, the elementary patterns are matched via the function graphMapping. This function returns the segments of the input DAG, which are rooted at the current node and matches with a given pattern. This step can be considered similar to the K-feasible cut generation of SMAC (for a K-LUT). The mapping solutions are generated via graph mapping against the pattern library. An exemplary graph mapping flow is illustrated with the Fig. 6.16. In this example, the node n_6 of the input graph is matched against the available patterns. A node is first matched with the root node of the pattern graph. A node-to-node matching is done by checking the operator (size, type) and the number of inputs. In the following iterations, the matched pattern graphs are traversed from root node upwards level-wise to check if those can completely cover a sub-graph of (I_d). Only in the case of complete cover, a pattern is considered to be a match for the current node. In the figure, several such matches (within rectangular boxes) are shown for pattern graphs P_1, P_2 and P_3 whereas no match is found for P_4.

The check with accommodation within predecessor nodes' clustering solutions actually resembles the clustering capacity check done in LUT-based FPGA synthesis.

Table 6.3 Data structure of arrival time table

Struct Arrival_Time_Row {	
Cluster_Type	m_cluster_solution;
Cluster_Id	m_cluster_id;
unsigned int	m_propagation_delay;
Bit_Matrix	m_bit_matrix_model;
List<Nodes>	m_node_list;};
Struct Arrival_Time_Table {	
List<Arrival_Time_Row>	m_arrival_time_row_list;};

A marked difference is that, for coarse-grained FPGA, the clustering capacity can
be defined in terms of the various elements available within the cluster as well as
the connectivity constraints allowed. To address this problem, an in-cluster place-
and-route is performed to check if the clustering capacity is met (presented in algo-

---

Algorithm 6.9: CG_SMAC_Labelling

Input: *graph_node_list, pattern_graphs, conn_constraint, fpga_topo, delay_model*
begin
    PI_node_list = sortLevelWise($graph\_node\_list$);
    foreach $node_{current}$ of PI_node_list **do**
        $mapped\_input\_graph\_segments$ = graphMapping($node_{current}$, $pattern\_graphs$);
        foreach $segment \in mapped\_input\_graph\_segments$ **do**
            $cluster_{new}$ = new Cluster($segment \rightarrow$ getClusterType(), getNewClusterID());
            $available\_elements$ = getAvailableElement($segment$, $cluster_{new}$, $fpga\_topo$);
            generateArrivalTimeRow($node_{current}$, $segment$, $cluster_{new}$, $available\_elements$, 0);
        $node_{current} \rightarrow min\_delay\_row\_list$ = getMinDelayInArrivalTimeTable($node_{current}$);

    foreach $node_{current} \in non\_PI\_node\_list$ **do**
        $mapped\_input\_graph\_segments$ = graphMapping($node_{current}$, $pattern\_graphs$);
        foreach $segment \in mapped\_input\_graph\_segments$ **do**
            $parent\_node\_list\_for\_segment$ = getParentNode($segment$);
            foreach $parent\_node \in parent\_node\_list\_for\_segment$ **do**
                $segment \rightarrow min\_delay\_list_{parent}$ .append($parent\_node \rightarrow min\_delay\_row\_list$);

            foreach $parent\_node \in parent\_node\_list\_for\_segment$ **do**
                $current\_arrival\_time\_table$ = $parent\_node \rightarrow$ getArrivalTimeTable();
                foreach $current\_arrival\_time\_row \in current\_arrival\_time\_table$ **do**
                    $cluster_{current}$ = $current\_arrival\_time\_row \rightarrow$ getCluster();
                    $available\_elements$ = getAvailableElement($segment$, $cluster_{current}$, $fpga\_topo$);
                    $is\_possible$ = inClusterPlaceAndRoute($cluster_{current}$, $segment$,
                    $available\_elements$, $conn\_constraint$);
                  if $is\_possible == true$ **then**
                      $prop\_delay$ = findPropDelay($segment$, $min\_delay\_list_{parent}$, $delay\_model$);
                      generateArrivalTimeRow($node_{current}$, $segment$, $cluster_{current}$,
                      $available\_elements$, $current\_arrival\_time\_row$);

            $cluster_{new}$ = new Cluster($segment \rightarrow$ getClusterType(), getNewClusterID());
            $available\_elements$ = getAvailableElement($segment$, $cluster_{new}$, $fpga\_topo$);
            $prop\_delay$ = findPropDelayWithNewCluster($segment$, $min\_delay\_list_{parent}$, $delay\_model$);
            generateArrivalTimeRow($node_{current}$, $segment$, $cluster_{new}$, $available\_elements$, $prop\_delay$);
        $node_{current} \rightarrow min\_delay\_row\_list$ = getMinDelayInArrivalTimeTable($node_{current}$);
end

Algorithm 6.10: inClusterPlaceAndRoute

Input: $cluster_{current}$, *segment, available_elements, conn_constraint*
Output: *is_routable*
begin
    $is\_routable$ = false;
    if $segment \subset available\_elements$ **then**
        $cluster_{current} \rightarrow$ append($segment$);
        min_cost = infinity;
        $possible\_placements$ = getAllPlacements($cluster_{current}$);
        foreach $placed\_i \in possible\_placements$ **do**
            $routed_i$ = route($placed\_i$, $conn\_constraint$, $is\_routable$);
            $routing\_cost$ = computeCost();
            if $is\_routable == true$ and $routing\_cost < min\_cost$ **then**
                min_cost = $routing\_cost$;
                $routed_{best}$ = $routed_i$;
        $segment \rightarrow$ appendClusterSolution($routed_{best}$);
    return $is\_routable$;
end

Algorithm 6.11: generateArrivalTimeRow

Input: $node_{current}$, $segment$, $cluster_{current}$, $available\_elments$, $prop\_delay$,
        $parent\_arrival\_time$

begin

    if $available\_elments \neq \emptyset$ **then**

        $Arrival\_Time\_Row$ $new\_arr\_time\_row$ = new $Arrival\_Time\_Row()$;

        if $parent\_arrival\_time$ **then**

            bit_matrix_model = Clone($parent\_arrival\_time \rightarrow$
            getBitMatrix());

        else

            $bit\_matrix\_model \rightarrow$ Initialize();

        bit_matrix_model.setBit($available\_elments \rightarrow$first());

        $new\_arr\_time\_row \rightarrow$ setCluster($cluster_{current}$);

        $new\_arr\_time\_row \rightarrow$ setPropDelay($prop\_delay$);

        $new\_arr\_time\_row \rightarrow$ setBitMatrixModel($bit_m atrix_m odel$);

        $new\_arr\_time\_row \rightarrow$ setNodeList($parent\_arrival\_time \rightarrow$
        getNodeList());

        $new\_arr\_time\_row \rightarrow$ getNodeList() $\rightarrow$ append($segment \rightarrow$
        getNodeList());

        $node_{current} \rightarrow$ appendArrivalTimeRow($new\_arr\_time\_row$);

end

---

rithm 6.10). The algorithm is run over each mapping solution of the current node. The key part of the algorithm is to decide if a predecessor node's existing clustering solution do have capacity to take the current node's mapping solution. This is done first by checking if the cluster has an unfilled logic block corresponding to the mapping solution. In that case, the mapping solution is added to form a new clus-

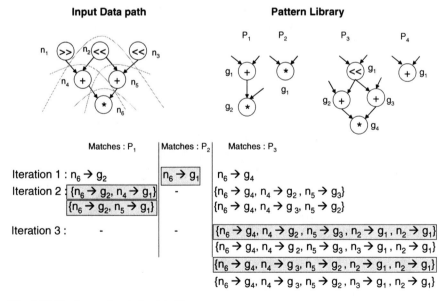

Fig. 6.16 Graph mapping during labelling

ter. However, this is not sufficient. Given the existing cluster's connectivity restrictions the newly added mapping solutions may not be routable. This is checked via generating all possible placement combinations within the scope of $cluster_{pred}$ and performing routing. Out of the various possible placements, the one with minimum routing cost is added to the current segment's possible clustering solutions. For each clustering solution, an arrival time row is constructed and tagged with the current node. This is done in the function generateArrivalTimeRow (presented in algorithm 6.11). The propagation delay at the entry of all the PI nodes are set to zero. Finally, the row(s) containing the minimum propagation delay (referred as $min\_delay\_row\_list$) are attached to the current node for quick retrieval.

After the initialization of the PI nodes with arrival time rows, the non-PI nodes of the input DAG are traversed level-wise towards the direction of Primary Outputs (POs). For all non-PI nodes, the patterns rooted at the node are matched, which corresponded to segments of the input DAG. For each of these segments, the parent nodes are determined. The $min\_delay\_row\_list$ for the parent nodes are collected. Then within the loop for each of the segments, the loop for parent nodes is triggered. For each of the arrival time rows of these parent nodes, an unique cluster solution does exist. The current node can be included in that cluster, given it can be fit in the rest of the available elements and it satisfies the connectivity constraints. Exactly these two are checked at the innermost loop of the non-PI nodes. If these conditions are met, a new arrival time row for the current node is created by mostly cloning the parent node's arrival time row. Few things are distinguished, though. Firstly, the bit matrix model of the parent node's arrival time row is modified to indicate that one more element is now occupied. Secondly, the propagation delay is modified by considering the current nodes intrinsic delay (D_{node}), the inter-cluster wire delay (D_{int}) and the maximum propagation delay among the parent nodes' minimum propagation delays. For the current node, the propagation delay is taken from the best routing solution (from algorithm 6.10). Lastly, the m_node_list member of parent node's arrival time row is extended by the current segment's list of nodes.

Following the check, even if the current segment can be fit into the existing clusters or not, a new cluster is also created. This new cluster begins from the current segment. Accordingly, the propagation delay is calculated. A corresponding entry is made to the current node's arrival time table. Proceeding in such way, the best arrival times for the PO nodes are finally calculated.

Cluster Realization : The cluster realization part of CG-SMAC is relatively straightforward. Commencing from the PO nodes, the algorithm executes until all the nodes are clustered i.e. until the set $nodes\_not\_clustered$ is empty. At each step, several nodes ($nodes\_just\_clustered$) are clustered, which are removed from the $nodes\_not\_clustered$. In the subsequent step, the parent nodes of the set $nodes\_just\_clustered$ are dealt with. For each node in each iteration, the arrival time row with best propagation delay is selected. There can be multiple rows fitting this criteria. Under such circumstance, the arrival time row providing the best cluster utilization is chosen. Better cluster utilization is marked - if a cluster consists of larger number of nodes in it. Since the segments from the input DAG are put inside a cluster, it is possible that both the parent and child nodes are included in the

Algorithm 6.12: CG_SMAC_Clustering

Input: *graph_node_list*
Output: *cluster_result*
begin

    cluster_result → initialize();
    PO_node_list = getPrimaryOutputs(*graph_node_list*);
    nodes_not_clustered = *graph_node_list*;
    current_level_nodes = *PO_node_list*;
    while *nodes_not_clustered* $\neq \emptyset$ **do**
        nodes_just_clustered → clear();
        while *node* = *current_level_nodes* → next() **do**
            min_delay_row_list = getMinDelayInArrivalTimeTable(*node*);
            Arrival_Time_Row best_arr_time_row =
            selectRowWithMaxUtilization(*min_delay_row_list*);
            cluster_new = new Cluster();
            cluster_new → getCopy(*best_arr_time_row* → getNodeList());
            cluster_result → append(*cluster_new*);
            nodes_just_clustered → append(*best_arr_time_row* →
            getNodeList());
            nodes_not_clustered → remove(*best_arr_time_row* →
            getNodeList());
            current_level_nodes → remove(*best_arr_time_row* →
            getNodeList()) sortLevelWise(*current_level_nodes*);
        current_level_nodes → clear();
        current_level_nodes → append(*nodes_just_clustered* →
        getParentNodes());
    return *cluster_result*;
end

current_level_nodes, thus being processed in one single iteration. To ensure that the child nodes (nodes nearer to the PO) are processed earlier and a single node is not processed twice, the *current_level_nodes* is updated and level-wise sorted at the end of each iteration. The algorithm proceeds in that way to realize the clustering solution (making mapping solutions along the way) with optimal propagation delay.

Complexity Analysis : Clearly, the runtime complexity of CG-SMAC is dictated by the cluster labelling phase. Let us consider an input boolean network of $\mathcal{N}$ nodes. For each such node, a matched pattern yields a segment. For a pattern library of $\mathcal{P}$ different elements, the number of segments is of the order $\mathcal{P}$, which also is of same order as far as the number of parent nodes for each segment is concerned. Every arrival time row for these parent nodes are searched for an available cluster in the function inClusterPlaceAndRoute (algorithm 6.10) inside the innermost for-loop of algorithm 6.9. The number of arrival time rows for a given node is a product of the number of arrival time rows of all its parent nodes (considering a possible clustering solution emerging from each one) and its possible pattern matches. A possible clustering solution from a parent node will only cease to be existent in case the cluster reaches its capacity, which is assumed to be of $\mathcal{C}$ elements. Additionally, a new arrival time row is created for each node. Therefore, for every node, the number of arrival time rows can grow up to $\mathcal{N}/\mathcal{L}^C*\mathcal{P} + 1$, where $\mathcal{L}$ is the number of levels in the input graph. The number of all possible placements inside a cluster of capacity $\mathcal{C}$ can grow up to $\mathcal{C}!$ in worst case scenario. However, usually a cluster is filled up with heterogeneous elements, reducing the complexity down. In case of say, h_1 and h_2 being the number of elements of two different types, the number of possible placement becomes $h_1!*h_2!$. Considering worst case scenario, the total time-complexity is therefore $\mathcal{O}(\mathcal{C}!*\mathcal{N}*\mathcal{P}^2*((\mathcal{N}/\mathcal{L})^\mathcal{C}*\mathcal{P} + 1))$, which is roughly equivalent to $\mathcal{O}(\mathcal{C}!*(\mathcal{N}/\mathcal{L})^{\mathcal{C}+1}*\mathcal{P}^3)$. For a definite pattern library, $\mathcal{P}^3$ can be considered to be a large constant factor. The clustering capacity, also, is usually low (≤ 6) [34, 186] for coarse-grained FPGAs.

Optimality Analysis : By the conservation of all the possible solutions under given structural constraints, it is guaranteed that the labelling phase determines the best arrival time for each node. This is true for only combinatorial networks, to which the CG-SMAC algorithm is applied in this work. While determining the clusters starting from PO nodes, it is, also, evident that the solution existing at the PO node is trace-able up to the PI nodes. This can only be possibly hampered if during labelling, overlapping solutions are created. The overlapping solutions can occur due to one node being mapped into multiple clusters or multiple patterns. However, by duplicating the overlapped nodes, the delay optimality can still be guaranteed.

6.10 FPGA Synthesis : Placement and Routing

Post mapping and clustering, the clustering solutions are to be placed and routed on the given FPGA structure. As stated previously, the placement solution needs to be routable. At the same time, the quality of routing depends strongly on the placement results. For this reason, the placement is performed with a measure of routability done within. Module placement is shown to be an NP-hard problem by Sahni and Bhatt in 1980 [187], which states that it is not possible to solve the placement problem in polynomial time by a deterministic Turing machine. To determine an exact solution of placement would mean evaluating all the cluster placement possibilities. For $\mathcal{C}$ clusters and $\mathcal{L}$ placement locations (with $\mathcal{L} \geq \mathcal{C}$), this would require time in the order of $\mathcal{P}_\mathcal{C}^\mathcal{L}$ (i.e. $\mathcal{L}!/(\mathcal{L} - \mathcal{C})!$). Evidently, that is not a possible solution of circuits with a reasonable number of clusters and placement locations. Several heuristic approaches, thus, emerged for solving the placement problem in ASIC and FPGA synthesis. The best results have been reported so far with simulated annealing-based heuristics [177]. In this book, the state-of-the-art algorithms placement algorithm [175] based on simulated annealing is applied. Within the core of this algorithm, a negotiation-based routing algorithm called *Pathfinder* [179] is utilized. Since, the routing algorithm forms a basic core of the placement algorithm, in the following the former is discussed first.

Routing : In plain words, the routing problem can be stated as to find if a mapped-and-placed set of modules are routable at all. If it is, then the goal is to connect all the modules while minimizing the critical path. The whole process of routing is fixedly driven by two constraints. The first one is the mapped-and-placed set of modules. This is considered to be un-alterable during the routing phase. The second constraint is the routing resources. Formally, the placed set of modules can be modelled as a directed multigraph $M_{placed} = \langle V, E \rangle$, where V is the set of vertices representing modules with their placement location. E is the multiset of edges connecting two vertices. Being a multiset, it is possible for E to have multiple edges between two vertices. With this, the routing problem can be formally defined as follows.

Problem Formulation :   Given a particular FPGA architecture with its topology definition T_{topo}, connectivity constraint definition $Conn_\mathcal{E}$ - determine the routing of a given input multigraph M_{placed} on the FPGA architecture - to minimize the critical path.

Solution Approach : The Pathfinder [179] routing algorithm is marked by two nested loops. The outer loop, referred as global router, is executed as long as any routing resource is congested. The inner loop, termed signal router, is called by the global router to completely re-route the signal (referred as *cluster_connections* henceforth) by avoiding obstacle. The routing of signal is performed by breadth-first search among the fan-outs of current *cluster* to get the lowest cost path. The cost of a route is a key point in the Pathfinder algorithm. The cost is computed by the usual path-delay as well as the routing resource cost. The cost of a routing resource increases with every iteration if it is congested. This slow increase in the routing resource sot forces the *cluster_connections* to avoid using highly over-used resources and thus find congestion free routing. At the same time, the global router calls the *cluster_connections* in the order of their decreasing slack during each iteration. This allows high priority *cluster_connections* to be routed first avoiding congestion.

In this book, the pathfinder algorithm is modified to fit the generic FPGA structure scenario. For the coarse-grained FPGA case, dedicated routing resources may exist in form of bus, crossbar or even a cluster itself can be cast as a routing resource. In the structural modelling of FPGA, a provision for all these are maintained. This necessitates a graph theoretic model of the FPGA structure including the routing resources. This is explained using the following Fig. 6.17.

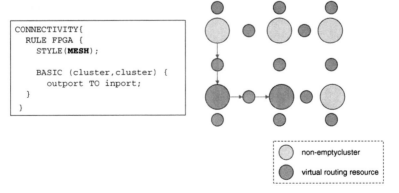

```
CONNECTIVITY{
   RULE FPGA {
      STYLE(MESH);

      BASIC (cluster,cluster) {
         outport TO inport;
      }
   }
}
```

○ non-emptycluster

● virtual routing resource

Fig. 6.17 Data-structure for Modelling Generic Routing Structure

In the Fig. 6.17, connectivity descriptions with the corresponding partial physical view of the FPGA is shown. Note that, only uni-directional edges are shown in the figure for the sake of clarity. In practice, for connecting between two adjacent clusters two virtual routing resources are used. As depicted in the Fig. 6.17, depending on the T_{topo}, a specific graph is built up to solve the routing problem. For different connectivity styles, different virtual routing resources are positioned around the clusters. Importantly, empty clusters/elements can be used as routing resources, as shown in the figure, if the cluster elements allow a direct path from the input to the output. In this case, the routing resources are dubbed *virtual* as no dedicated routing resources are modelled. Dedicated routing resources are also feasible to model using FPGA structural description, as also shown in the Fig. 6.17. Formally, the *clusters* and the routing resources are modelled as a directed multi-graph with two different types of vertices namely, clusters and routing resources. The *cluster_connections* are routed from the source to sink using pathfinder algorithm employing these routing data-structure.

Placement : Informally, the placement problem can be defined as - to find positions for placing all the mapped clusters on the given FPGA structure to minimize the overall routing cost. Let us consider that there are C mapped clusters and $\mathcal{L}$ locations for these clusters on the FPGA. Let S be the set of N_s *cluster_connections*, with W_i ($i \leq N_s$) being the weight of each *cluster_connection*. A mapping function F(i, j) can be defined to be 1 if i-th *cluster* is placed in j-th position, 0 otherwise. Therefore the cost function can be formulated to be $W_i \cdot$F, which denotes the weight of each *cluster_connection* under a particular placement assignment. On that basis the placement problem can be formulated as following. The given mapped cluster can also be represented using a multigraph as like a placed one. This time for $M_{\text{clustered}} = \langle V, E \rangle$, the vertices are not tagged with their placement position.

Problem Formulation : Given a particular FPGA architecture with its topology definition T_{topo}, connectivity constraint definition $Conn_\mathcal{E}$ - determine the placement of a given mapped cluster multigraph $M_{\text{clustered}}$ on the FPGA architecture - such that the following conditions are held.

- $$\sum_{j=1}^{\mathcal{L}} F(i, j) = 1$$

- $$\sum_{i=1}^{C} F(i, j) \leq 1$$

and the cost function $\sum_{i=1}^{N_s} W_i \cdot$F is minimized.

In the above problem formulation, the constraints ensure that every *cluster* is placed and that one *cluster* is placed only once. The cost function can take different forms depending on the weight measure of the *cluster_connections*. For *area-driven placement*, the density of channels are considered as weights, for *routability-driven placement* the total wire length is the weight measure, whereas for *timing-driven placement* the length of the critical path is taken as weight. The placement algorithm adopted in this book is termed as Coarse-grained Generic Simulated Annealing-based Placement or CG-SA-Placement. It is presented using pseudo-code in the algorithm 6.13. Note that, the area-optimized placement problem can be formulated as a bin-packing problem, which is known to be NP-hard.

The inputs of the CG-SA-Placement algorithm are the clustered input graph (i.e. output of CG-SMAC), the connectivity constraints, the FPGA structure topology and the delay model. At the very beginning of the algorithm, a new list of clusters, called *placed_cluster*, is created. This list is initialized by putting all the available clusters in the FPGA topology. Next to that, the clusters from CG-SMAC output are placed with random co-ordinates in this list. The other clusters remain empty at this point. Before starting with the simulated annealing, the starting temperature is determined. This is done via maze-routing the connections, swapping cluster positions several times, calculating the placement costs and summing it up. The initial temperature is set to be a multiple of the standard deviation of these placement costs. The starting temperature, the cooling schedule of simulated annealing and the terminating temperature (which is equal to $0.005 * placement\_cost$ / # *cluster_connections*) are set according to that of the state-of-the-art fine-grained FPGA place-and-route tool [176].

Within each temperature iterations, a number of movements are allowed. For each movement, two particular clusters from the *placed_cluster* are chosen. For these two clusters the associated connections are completely ripped up. After swapping, the new connections are established by maze routing. Followed by that, the pathfinder routing algorithm is called upon. Depending on the pathfinder outcome, the new placement cost is computed. The movement is accepted or not is based on the current annealing temperature and the difference of costs between earlier placement (before swapping) and the current one. If the movement is accepted then the current routing path is cached. If it is not, then the clusters are swapped back to their original positions. The old routing path is also re-established from the previously cached routing path.

6.10.1 Configuration Bitstream Generation

Generating the configuration bitstream out of the placed-and-routed FPGA is a methodical task. For this work to be generic, the configuration bits necessary at each point or arbitration is directly taken from the FPGA-IR. Surely, for a different FPGA structural definition, the configuration bits and their distribution becomes different.

Algorithm 6.13: CG_SA_Placement

Input: *cluster_smac, conn_constraint, fpga_topo, delay_model*
Output: *placed_cluster*
begin

    placed_cluster → Initialize(*fpga_topo*);
    foreach *cluster$_{current}$* ∈ *cluster_smac* **do**
        cluster$_{current}$ → setRandomCoordinate(*fpga_topo*);
        placed_cluster → setCluster(*cluster$_{current}$*);
    cluster_connection_list = routeMaze(*placed_cluster*);
    placement_cost$_{current}$ = computeCost(*cluster_connection_list, delay_model*);
    placement_cost_list → append(*placement_cost$_{current}$*);
    unsignedint cluster_count = *placed_cluster* → getCount();
    for *i = 1* **to** *cluster_count* **do**
        cluster$_A$ = randomChooseCluster(*placed_cluster*);
        cluster$_B$ = randomChooseCluster(*placed_cluster*);
        swapCluster(*cluster$_A$, cluster$_B$*);
        cluster_connection_list = routeMaze(*placed_cluster*);
        placement_cost$_{current}$ = computeCost(*cluster_connection_list, delay_model*);
        placement_cost_list → append(*placement_cost$_{current}$*);
    temperature = computeInitialTemperature(*placement_cost_list*);
    placement_cost$_{old}$ = *placement_cost$_{current}$*;
    cacheRoutingPath(*cluster_connection_list*);
    while *temperature > terminating_temperature* **do**
        for *i = 1* **to** *movement_count* **do**
            cluster$_A$ = randomChooseCluster(*placed_cluster*);
            cluster$_B$ = randomChooseCluster(*placed_cluster*);
            ripped_up_connection_list = ripUpConnection(*cluster$_A$, cluster$_B$*);
            swapCluster(*cluster$_A$, cluster$_B$*);
            foreach *connection* ∈ *ripped_up_connection_list* **do**
            routeMaze(*connection*);
            PathFinder(*cluster_connection_list, cluster_smac* → getPrimaryInput(), *cluster_smac* → getPrimaryOutput());
            placement_cost$_{new}$ = computeCost(*cluster_connection_list, delay_model*);
            δ_{cost} = *placement_cost$_{new}$* - *placement_cost$_{old}$*;
            if acceptMovement(δ_{cost}, *temperature*) == *true* **then**
                placement_cost$_{old}$ = *placement_cost$_{new}$*;
                cacheRoutingPath(*cluster_connection_list*);
            else
                swapCluster(*cluster$_A$, cluster$_B$*);
                copyCachedRoutingPath(*cluster_connection_list*);
        temperature → update();
    return *placed_cluster*;
end

During the FPGA-IR construction, these bits are allocated as elaborated in the previous *connectConfiguration* function (refer algorithm 6.8).

The function to generate the configuration bitstream accepts the *placed_cluster* and the configuration information from the FPGA-IR as input. The configuration information contains the name of each FIR-Port for each FIR-Entity. For each FIR-Port, the configuration bits to control it (if any) with their relative bit-position in the top-level configuration bitstream are stored. Furthermore, the exact bitstream added with the corresponding target FIR-Port completes the configuration

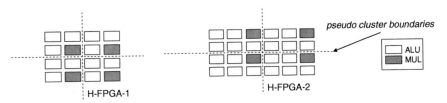

Fig. 6.18 Heterogeneous FPGAs

information. Exemplarily, say a given FIR-Port `clusteroutport` can connect to `clusterinport` of its four neighbouring clusters. The configuration information stored at the FIR-Port `clusteroutport` will thus contain bit-positions (say 32..31) to control it. It will also contain the target FIR-Ports with the corresponding control bit-streams (e.g. 00 → clusterinport_cluster1, 01 → clusterinport_cluster2, 10 → clusterinport_cluster3, 11 → clusterinport_cluster4). Similar pieces of information are stored for the configuration of FIR-Processes (containing OPERATOR_LIST), too. With the *placed_cluster* containing detailed information about the connected ports, the bitstream generation task only iterates among the FIR-Entities recursively, thereby setting the corresponding configuration bits during the process.

6.10.2 Synthesis on Non-Clustered Heterogeneous FPGA

The proposed FPGA description style permits modelling of a heterogeneous FPGA by allowing various elements to be grouped in a cluster and various different clusters forming the complete FPGA. However, adapting CG-SMAC and the placement-and-routing algorithm for heterogeneous FPGA is a major challenge as those are designed with clustered LUT-based FPGAs in view. One possible alternative is to model the entire heterogeneous FPGA as one single large cluster. This will pose difficulty to the in-cluster placement and routing phase, which permutes over all possible placement positions. In this work, the chosen alternative is to determine some form of homogeneity within the heterogeneous FPGAs and thereby, impose *pseudo cluster boundaries*. In this case, the inter-cluster routing cost is considered equal to the intra-cluster one. This enables the entire FPGA synthesis flow presented so far to be applied to heterogeneous FPGAs. Exemplary FPGAs, with pseudo cluster boundaries are shown in the Fig. 8.6. Surely, the homogeneity of the so called heterogeneous FPGAs can be more difficult to find in some cases, leading to the formation of a single large cluster and thereby increasing the complexity of synthesis algorithms. However, it can be argued that, increasing heterogeneity makes the FPGA more ASIC-like and thus, it should be natural to expect the synthesis algorithms to run for longer time then.

6.11 Synopsis

- The result of the pre-fabrication implementation is an RTL description of the rASIP completed with synthesis, simulation scripts as well as the configuration bit-stream for the FPGA.
- Pre-fabrication implementation does not merely follow the choices of pre-fabrication exploration but, also allows a larger exploration loop by providing more detailed feedback on the earlier decisions. This calls for a flexible implementation flow as elaborated in this chapter.
- The implementation flow is rooted in an ASIP design environment. By high-level keywords it is possible to partition the processor in fixed and re-configurable block. Furthermore, the structural details of the re-configurable block can also be designed using an abstract description style.
- The pre-fabrication implementation flow can be grossly put into three phases namely, base processor implementation and partitioning, FPGA implementation and FPGA synthesis.
- The genericity as well as the implementation flexibility strongly relies on the algorithms (adopted, designed or modified) integrated in the flow.

Chapter 7
Post-fabrication Design Space Exploration and Implementation

In der Beschränkung zeigt sich erst der Meister.
Goethe, Polymath, 1749–1832

The designer enters post-fabrication phase of rASIP design, when most of the design decisions are already taken. This phase is important, therefore, to make the optimum decisions within the constraints set in the pre-fabrication phase. The complete tool-flow specified in the pre-fabrication phase remains same in post-fabrication phase, too. In the same manner as before, one has to perform profiling of new, evolving applications, followed by instruction-set simulation, RTL simulation etc. The difference is that all these are done in a constrained manner. For example, after profiling the application - the designer is allowed to change only a part of the LISA model namely, the part targeted for re-configurable block. Therefore, the post-fabrication design space exploration can be alternatively termed as *constrained design space exploration*. The pre-fabrication design space exploration is, contrarily, *unconstrained design space exploration*.

The task of post-fabrication design space exploration is basically to select between alternative application partitions, which will be mapped to the re-configurable portion and the fixed portion of the rASIP. Considering that the entire application is first targeted to the base processor, the design space exploration is simply to select a basic block from the application, form a special custom instruction out of it using LISA 3.0 description, map it to the coarse-grained FPGA and evaluate the results. Though this entire flow can be considered as a large design space exploration yet, due to the different nature of problems addressed, the selection of custom instruction is covered within post-fabrication design space exploration. The later steps namely, HDL generation for re-configurable part and mapping of those to the coarse-grained FPGA is discussed in the later Section 7.2 elaborating post-fabrication design space exploration.

7.1 Post-fabrication Design Space Exploration

The post-fabrication design space exploration flow is shown in the Fig. 7.1. The black boxes shown in the figure refers to the design components, which are fixed in

A. Chattopadhyay et al., *Language-driven Exploration and Implementation of Partially*   139
Re-configurable ASIPs, DOI 10.1007/978-1-4020-9297-8_7,
© Springer Science+Business Media B.V. 2009

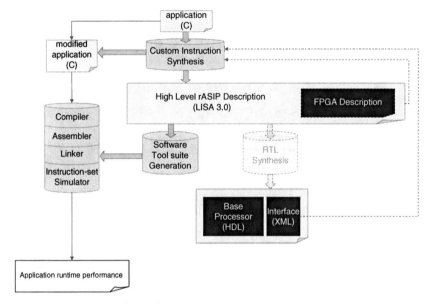

Fig. 7.1 rASIP post-fabrication design space exploration

the post-fabrication phase. These restricts the design space of the custom instruction synthesis tool. Thus, during the custom instruction selection, hard constraints need to be met. In the following, all such constraints are listed. Some of these constraints are directly fed into the custom instruction synthesis tool and some of these are verified manually.

- The special-purpose or custom instructions must not violate the existing interface between re-configurable block and the base processor.
- The additional number of instructions must not exceed the number allowed by the free opcode space available.
- The custom instructions selected, must be within the area-budget of the coarse-grained FPGA.

7.1.1 Integration with Custom Instruction Synthesis Tool

The state-of-the-art custom instruction synthesis tool, named Instruction Set Extension Generator (ISEGen) [113] is integrated with the rASIP tool-flow proposed in this work. The integration proves to be particularly useful during the post-fabrication rASIP design space exploration since, a large number of design decisions are already taken at this point. Thus, the point of automation and optimization in the tool-flow is much clearly established. The ISEGen fits this perfectly. The inputs to the ISEGen are firstly, the application or part of it and secondly, the design constraints. The outputs are firstly, the modified application with special inline assembly functions replacing plain C code. Secondly, the declaration of inline assembly functions so that, the application can be compiled. Finally, the definition of the inline assembly functions in plain C is produced by ISEGen. The definition of the inline assembly functions needs to be ported as LISA description with corresponding opcode and syntax definition. Nevertheless, the C description produced as an output of ISEGen is used effectively for host-based debugging purposes.

Feeding ISEGen Input : ISEGen offers a GUI-based frontend for editing and/or selecting a part of the application, which the designer wants to choose custom instruction from. The selected part is then subjected to an Integer Linear Programming (ILP)-based algorithmic flow to obtain the custom instructions. The criteria for identifying a custom instruction includes several clauses. Exemplarily, the identified data-flow graph should be *convex* i.e. it must not have one external node both receiving and sending data from/to it. On top of these basic analysis, the constraints from pre-fabricated rASIP design are fed to the ISEGen. The constraints are basically of two types. Firstly, the *interface constraints* and secondly, the *structural constraints*. While interface constraints are about the availability of connecting ports between the base processor and the re-configurable block, the structural constraints are about selecting the custom instructions to fit the pre-defined coarse-grained FPGA perfectly.

The first set of constraints (stored in the so called `config.xml` file) deals about the interface restrictions. Inside this, the number of General Purpose Register (GPR) read-write ports available to the custom instructions are stored. Furthermore, the number of memory ports and number of local scratchpad memories with their port count is stored. Localized scratchpad memories inside the re-configurable block can boost performance as shown in [115, 188]. The parameterizability of scratchpads offers an unique advantage of ISEGen to explore that fully. Finally, subsequently occurring custom instructions may need to store intermediate results locally. This can be achieved by having a local register file. The number of ports allowed for each custom instruction to/from the local register file is also specified in the interface constraints.

The second set of constraints, which is stored in the `param.xml` file, holds several details of the coarse-grained FPGA structure. This may include the collection of basic elements e.g. LUT, MULT, ADD available in the coarse-grained structure. Further details like, the bit width of these basic operators and the count of each such operator can also be stored. A future enhancement of this tool will allow few connectivity details to be incorporated in the constraint-set as well. This will mean a high-level mapping and clustering during the custom instruction selection itself. Note that, exactly this is the point where the custom instruction selection, mapping, placement and routing are merged together by several other tools [33, 34]. In contrast, the decoupled flow suggested in this book [189] allows more design points to be taken into consideration.

Exporting ISEGen Output : The modified application and the inline assembly functions' declaration, generated by ISEGen, are used as it is for the rASIP post-fabrication design space exploration. The generated definition of inline assembly functions do need some follow-up work in order to put that as a LISA 3.0 description. The ISEGen produces the behavior definition of custom instruction in XML format. A software program named *ISE2LISA* is developed to automatically read the XML description and write a LISA description out of it. The translation of behavior portion to LISA behavior section is almost verbatim except for the the following macros used by ISEGen.

- UDI_RS and UDI_RT : These macros are used to read from the general purpose registers. In ISE2LISA, these is transformed to the R[UDI_RS]; and R[UDI_RS]; statements respectively, where R is the general purpose register in the resource section of given LISA model.
- UDI_WRITE_GPR(value) : This macro is used to write to the general purpose register. This macro is transformed to R[UDI_WRITE_GPR] = value;, where R is the general purpose register.
- SCRATCH_MEM : This macro is used to read/write from/to the local scratchpad memory. This has to take into account the memory access identifier as well as synchronous, asynchronous nature of the memories into account. The translation of this is currently done manually.

Apart from the behavior part, the LISA definition of an instruction needs to consist of syntax, opcode and its location in the overall LISA operation graph. The location of the newly determined custom instructions in the overall LISA operation graph is already fixed in the pre-fabrication phase. The opcode of the custom instruction can be pre-determined if the decoder is not localized. Otherwise, the opcode is determined and allocated by using the *Coding Leakage Explorer* tool. If the instruction is accessing GPRs, then those also need to have their placeholders in the instruction syntax and instruction opcode. These are determined by the designer's knowledge of GPR opcode and syntax, which are fixed during the post-fabrication phase. Once the mapping of custom instructions' behaviors are done, the custom instructions are added to the UNIT definition of LISA model.

7.1.2 Re-targeting Software Toolsuite

After the enhancements done in the LISA description, the software tools need to be re-generated. This task is fairly simple for the simulator, assembler, linker and loader. As already discussed previously, the only tool deserving special attention among the above is the simulator. That too, when an instruction with the keyword *latency* is used. Contrarily the task of re-targeting the C compiler requires extra efforts. Earlier in chapter 5, the re-targeting of simulator under special cases and the C compiler are described. The same flow is also used here.

7.2 Post-fabrication Design Implementation

The post-fabrication design implementation flow, proposed in this book, is as shown in the Fig. 7.2. The post-fabrication implementation part concentrates on primarily

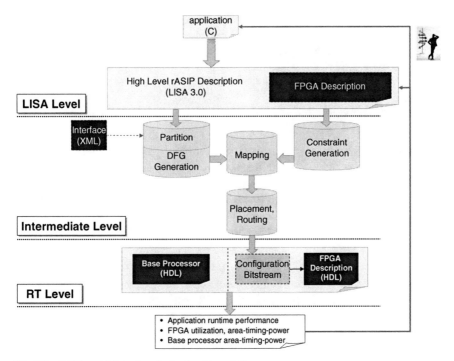

Fig. 7.2 rASIP post-fabrication design implementation

two things. Firstly, to ensure that the interface restrictions from pre-fabrication design decision is strictly maintained during implementation. This is actually loosely imposed in the post-fabrication exploration loop. Secondly, to map, place and route the datapath selected for re-configurable block on to the coarse-grained FPGA. In the following sub-sections, these issues are elaborated.

7.2.1 Interface Matching

The interface between the base processor and the re-configurable part is automatically generated during IR generation phase of RTL synthesis from LISA. The generated interface is stored internally and the earlier stored interface (in .xml format) is retrieved. The interface matching is performed, followed by the port allocation. The interface matching is done as presented in algorithm 7.1. For brevity, the matching of only some interesting IR-paths are covered in the pseudo-code.

The currently generated interface is stored in form of IR-paths internally. For different kinds of IR-paths, various unique attributes are stored. These attributes are matched against the corresponding IR-paths in the pre-fabrication interface. For example, the IR-path activation (representing a signal triggering an operation) is unique by its direction, signal name and the name of the operation it is triggering.

These are matched as shown in the algorithm. It is important to note that the interface matching does not necessarily produce an one-to-one solution in some cases e.g. for matching of a register read-write port. There can be several register read-write ports existing in the pre-fabrication interface. The newly generated interface may use one of them to access the register file. Therefore, during this process the first match is selected.

Three different kinds of errors can occur during interface matching.

Unmatched Interface This error is faced when a particular newly generated interface is not existing in the pre-fabrication interface at all.

Under-used Interface This is more of a warning than an error, triggered when all the pre-fabrication interfaces are not used.

Over-used Interface This is an error faced, when the sheer number of required interfaces in the post-fabrication mode exceeds that set up during pre-fabrication.

Interestingly, any of the three aforementioned errors can occur together with the other ones. For example, while a particular IR-path activation signal can get un-matched, it produces an under-used as well as an over-used interface directly. While a single register read-port can be left under-used. To enable the designer analyze the situation clearly, a tabular display is done after the HDL generation. The display shows all the previous interfacing signals and the corresponding matched signals. Additionally, it shows the interfaces which are not used as well as the newly generated unused interfaces, if any.

7.2.2 Area-Optimized Coarse-Grained FPGA Synthesis

During the post-fabrication phase, the dimensions of the coarse-grained FPGA remains completely fixed. Therefore, the main objective of coarse-grained FPGA synthesis stage is to fit the given datapath within it. Keeping this in perspective, in post-fabrication implementation phase - area optimization is investigated within the algorithmic steps of coarse-grained FPGA synthesis.

As explained in the previous chapter, the CG-SMAC algorithm accomplishes mapping and clustering with optimum delay as goal. CG-SMAC algorithm has two main components. First, the labelling phase and second the cluster realization phase. During labelling phase, the nodes of input datapath are labelled with all possible arrival times starting from PI nodes down to PO nodes. During cluster realization phase, the traversal begins from PO nodes and the best possible arrival time at each node is chosen. The same principle of operation is used for implementing area-optimized CG-SMAC. In general, obtaining the area-optimal solution is of exponential complexity and therefore not attempted. A simple first-fit heuristic is used. The algorithmic steps for labelling in area-optimized CG-SMAC are exactly same as regular CG-SMAC. The major differentiator is in the generation of arrival

---

Algorithm 7.1: matchInterface

Input: $list_{ir\_path\_prefab}$, $list_{ir\_path\_current}$

begin

 $IR\_Path\ ir\_path_{prefab}$;
 $IR\_Path\ ir\_path_{current}$;
 foreach $ir\_path_{current} \in list_{ir\_path\_current}$ **do**
  foreach $ir\_path_{prefab} \in list_{ir\_path\_prefab}$ **do**
   if ir_path_{prefab}.getType $()$ == $ir\_path_{current}$.getType $()$ **then**
    if ir_path_{prefab}.getType $()$ == $IR\_Path\_Register$ **then**
     if ir_path_{prefab}.getDirection $()$ == $ir\_path_{current}$.getDirection $()$
     and ir_path_{prefab}.getName $()$ == $ir\_path_{current}$.getName $()$
     and ir_path_{prefab}.getDatatype $()$ == $ir\_path_{current}$.getDatatype $()$
     and ir_path_{prefab}.getReadOrWritten $()$ ==
     $ir\_path_{current}$.getReadOrWritten $()$ **then**
      $ir\_path_{current}$.matched = **true**;
      $ir\_path_{current}$.match_port = ir_path_{prefab};
      $list_{ir\_path\_prefab}$.remove (ir_path_{prefab});

    else if ir_path_{prefab}.getType $()$ == $IR\_Path\_Activation$ **then**
     if ir_path_{prefab}.getDirection $()$ == $ir\_path_{current}$.getDirection $()$
     and ir_path_{prefab}.getName $()$ == $ir\_path_{current}$.getName $()$
     and ir_path_{prefab}.getDatatype $()$ == $ir\_path_{current}$.getDatatype $()$
     and ir_path_{prefab}.getOperationName $()$ ==
     $ir\_path_{current}$.getOperationName $()$ **then**
      $ir\_path_{current}$.matched = **true**;
      $ir\_path_{current}$.match_port = ir_path_{prefab};
      $list_{ir\_path\_prefab}$.remove (ir_path_{prefab});

    else if ir_path_{prefab}.getType $()$ == IR_Path_Coding **then**
     if ir_path_{prefab}.getDirection $()$ == $ir\_path_{current}$.getDirection $()$
     and ir_path_{prefab}.getName $()$ == $ir\_path_{current}$.getName $()$
     and ir_path_{prefab}.getDatatype $()$ == $ir\_path_{current}$.getDatatype $()$
     and ir_path_{prefab}.getContext $()$ == $ir\_path_{current}$.getContext $()$ **then**
      $ir\_path_{current}$.matched = **true**;
      $ir\_path_{current}$.match_port = ir_path_{prefab};
      $list_{ir\_path\_prefab}$.remove (ir_path_{prefab});

end

---

time data-structures, where additional entries are made to keep a count of number of clusters allocated along each path from PI to PO. The modified algorithmic steps for generation of arrival time row are shown in the algorithm 7.2. From the algorithm, it can be observed that a new entry named *cluster_count* is added to the arrival time table data-structure. In case, there is a *parent_arrival_time* i.e. the current node is accommodated in an existing cluster, then the arrival time row's cluster count is not incremented. Otherwise, it is increased by 1. By this way of labelling, every node remembers the number of clusters it currently consumes.

The cluster realization phase selects the mapping and clustering option with minimum cluster count along each path from PO to PI. This is elaborated via pseudo code in the following algorithm 7.3. The algorithm works in the same manner as in delay-optimal version except that the arrival time row with minimum cluster count is chosen.

Algorithm 7.2: generateArrivalTimeRow

Input: $node_{current}$, $segment$, $cluster_{current}$, $available\_elments$, $prop\_delay$, $parent\_arrival\_time$
begin

    if $available\_elments \neq \emptyset$ **then**
        $Arrival\_Time\_Row$ $new\_arr\_time\_row$ = new $Arrival\_Time\_Row()$;
        if $parent\_arrival\_time$ **then**
            bit_matrix_model = $\texttt{Clone}$ ($parent\_arrival\_time \rightarrow \texttt{getBitMatrix}()$);
            $new\_arr\_time\_row \rightarrow cluster\_count = parent\_arrival\_time \rightarrow cluster\_count$;
        else
            $bit\_matrix\_model \rightarrow \texttt{Initialize}()$;
            $new\_arr\_time\_row \rightarrow cluster\_count = parent\_arrival\_time \rightarrow cluster\_count + 1$;
        $bit\_matrix\_model.\texttt{setBit}$ ($available\_elments \rightarrow \texttt{first}()$);
        $new\_arr\_time\_row \rightarrow \texttt{setCluster}(cluster_{current})$;
        $new\_arr\_time\_row \rightarrow \texttt{setPropDelay}(prop\_delay)$;
        $new\_arr\_time\_row \rightarrow \texttt{setBitMatrixModel}(bit_matrix_model)$;
        $new\_arr\_time\_row \rightarrow \texttt{setNodeList}(parent\_arrival\_time \rightarrow \texttt{getNodeList}())$;
        $new\_arr\_time\_row \rightarrow \texttt{getNodeList}() \rightarrow \texttt{append}(segment \rightarrow \texttt{getNodeList}())$;
        $node_{current} \rightarrow \texttt{appendArrivalTimeRow}(new\_arr\_time\_row)$;

end

Algorithm 7.3: CG_SMAC_Clustering

Input: $graph\_node\_list$
Output: $cluster\_result$
begin

    $cluster\_result \rightarrow \texttt{initialize}()$;
    $PO\_node\_list = \texttt{getPrimaryOutputs}(graph\_node\_list)$;
    $nodes\_not\_clustered = graph\_node\_list$;
    $current\_level\_nodes = PO\_node\_list$;
    while $nodes\_not\_clustered \neq \emptyset$ **do**
        $nodes\_just\_clustered \rightarrow \texttt{clear}()$;
        while $node = current\_level\_nodes \rightarrow \texttt{next}()$ **do**
            $min\_cluster\_row\_list = \texttt{getMinClusterCountInArrivalTimeTable}(node)$;
            $Arrival\_Time\_Row$ $best\_arr\_time\_row =$
            $\texttt{selectRowWithMaxUtilization}(min\_cluster\_row\_list)$;
            $cluster_{new} = \texttt{new Cluster}()$;
            $cluster_{new} \rightarrow \texttt{getCopy}(best\_arr\_time\_row \rightarrow \texttt{getNodeList}())$;
            $cluster\_result \rightarrow \texttt{append}(cluster_{new})$;
            $nodes\_just\_clustered \rightarrow \texttt{append}(best\_arr\_time\_row \rightarrow \texttt{getNodeList}())$;
            $nodes\_not\_clustered \rightarrow \texttt{remove}(best\_arr\_time\_row \rightarrow \texttt{getNodeList}())$;
            $current\_level\_nodes \rightarrow \texttt{remove}(best\_arr\_time\_row \rightarrow \texttt{getNodeList}())$
            $\texttt{sortLevelWise}(current\_level\_nodes)$;
        $current\_level\_nodes \rightarrow \texttt{clear}()$;
        $current\_level\_nodes \rightarrow \texttt{append}(nodes\_just\_clustered \rightarrow \texttt{getParentNodes}())$;
    return $cluster\_result$;

end

7.3 Synopsis

- The post-fabrication exploration and implementation phase essentially utilizes the same tool-flow as in the pre-fabrication phase. The major difference is that several specifications (or parts of those) are frozen in the post-fabrication phase.
- With tighter constraints in the post-fabrication phase, it is important to apply those constraints on the exploration and implementation tools as well.
- A custom instruction synthesis tool is coupled in the post-fabrication design space exploration phase. The tool, termed ISEGen, determines potential special-purpose instructions and evaluates their impact on the overall application

performance with the help of LISA instruction-set simulator. Most importantly, this tool accepts various design constraints as existent in the post-fabrication flow.
- To ensure the constraints are not violated, an interface matching is performed during HDL generation from LISA.
- With the dimension restrictions of coarse-grained FPGA becoming hard in the post-fabrication phase, an area-optimized algorithm for synthesis is proposed.

Chapter 8
Case Study

Thinking is easy, acting is difficult, and to put one's thoughts into action is the most difficult thing in the world.

Goethe, Polymath, 1749–1832

8.1 Introduction

The tools developed during the course of this work and presented in the previous chapters are put to test via several experimental case studies. The empirical results obtained via those, created impetus for further research in various directions. Given the immense scope of the rASIP tools, it is not trivial to put them to work simultaneously, even though the guiding principle of application-driven processor design is maintained. There might be cases where one tool appears to need further maturity, whereas efficacy of the some other tool is firmly established. Some language extensions prove to be solidly required, while some other extensions are missed. The following case studies provide insight to the rASIP design framework with its chronological development.

The first two case studies demonstrate the experiments with two different classes of architectures namely, simple-scalar RISC and VLIW architectures. Different classes of applications, from the domain of cryptography, multimedia and signal processing, are chosen to drive the rASIP design. The third case study focussed solely on the exploration of coarse-grained FPGA architectures. In the fourth case study, a rASIP is built step-by-step for a baseband signal processing algorithm. In the final case study, the design exploration efficiency is demonstrated by tweaking at various phases of the tool flow and quickly observing its effect on other phases. For all the case studies, the rASIP architecture exploration started with some driver applications and an initial architecture.

8.2 Experiments with RISC-Based Architecture

In this section, first the target processor architecture is discussed. This is followed by a brief introduction of the chosen applications and then, the detailed elaboration of the case study.

The initial template architecture, LT_RISC_32p5, is a 32-bit, 5-stage pipelined architecture. The processor contains 16 general purpose registers and several

A. Chattopadhyay et al., *Language-driven Exploration and Implementation of Partially Re-configurable ASIPs,* DOI 10.1007/978-1-4020-9297-8_8, © Springer Science+Business Media B.V. 2009

special-purpose registers. LT_RISC_32p5 employs general purpose load-store in-structions, arithmetic instructions and instructions to perform boolean operations. The architecture also is interlocked and supports bypass mechanism.

The target applications for the rASIP are chosen from the domain of cryptogra-phy. These are listed in the following.

Blowfish Blowfish is a 64-bit block sized, 32–448-bit key length, symmetric block cipher. It is a free en/decryption algorithm, one of the fastest block-cyphers in wide-spread use today. It has two parts: key expansion and data encryption. In the key expansion phase, the variable-length key is used to generate a set of 32 bit sub-keys which are stored in arrays known as S-Boxes and P-Boxes. The encryption algorithm applies P-Box dependent permutation and S-Box dependent substitution on the plain text in 16 rounds. Current cryptanalysis could break only up to 4 rounds of the encryption.

DES Despite proven theoretical and analytical weaknesses, Data Encryption Stan-dard (DES) remains to be the most commonly used encryption algorithm for research purposes. This is due to the active involvement of government agencies in designing, using and further declassification of DES. The key-size of DES is 7 bytes. The algorithm works in 16 rounds i.e. processing stages. Furthermore, there is an initial and a final permutation round. The initial and final rounds are included deliberately to slow down the software execution of DES as it was designed for hardware. This, presumably, defended it from cryptanalysis at the time it was designed (early 1970s).

GOST GOST is a 64-bit block sized, 256-bit key length symmetric block cypher. Like Blowfish, GOST also uses S-Box substitution inside a similar F func-tion. GOST uses 8 S-Boxes and applies 32 encryption rounds on the plain text. [1]

As proposed in the design flow, the case study is divided into two phases i.e. the pre-fabrication and post-fabrication phases. The case study starts with the anal-ysis of applications for rASIP design in pre-fabrication phase and then in the post fabrication phase uses the flexibility offered by re-configurable block to optimize the rASIP. While, *DES* is used to take the pre-fabrication architectural decisions, *Blowfish* and *GOST* are used as the post-fabrication applications with the interface and the base processor remaining fixed.

Pre-fabrication Design Space Exploration: The application DES is chosen as the starting point for rASIP design. The application is subjected to the application profiling tool [105] for identification of hot spots. The function *des_round* is iden-tified as the hot spot of the DES application. This function is then subjected to the ISEGen tool for identification of custom instructions with initial constraints of a 2-input interface to 2 general purpose registers of the base processor, a 1-output interface to 1 GPR of the base processor and 32 internal registers, each of 32 bit

---

[1] This algorithm is the Russian counterpart of the American DES algorithm.

Table 8.1   Pre-fabrication synthesis results for LT_RISC_32p5

Re-configurable Block			Base Processor	
Area	Minimum Clock Delay (ns)	Latency	Area (Gates)	Clock Delay (ns)
1653 LUTs 512 Registers	12.09	4	88453	4.0

width. The ISEGen tool generated 5 custom instructions, the behaviors of which are mapped completely in LISA description. The codings of these custom instructions are determined using the *Coding Leakage Explorer*.

During RTL synthesis with these custom instructions, register localization and decoder localization options are turned on, in order to have flexibility for adding further custom instructions in the re-configurable part. The base processor is synthesized with Synopsys Design Compiler [139] using 0.13 μm process technology of 1.2 V. The re-configurable block is synthesized with Synopsys FPGA Compiler [190] using Xilinx Virtex-II pro [191] (0.13 μm process technology of 1.5 V) as the target device. The synthesis results are given in the Table 8.1.

As the synthesis results demand the FPGA latency to be at least 4 times that of the base processor, a latency of 4 is introduced during the simulation of the modified DES application. It is observed that the latency of custom instructions can be completely hidden by performing efficient manual scheduling. The initial simulation results of the DES application show up to 3.5 times runtime speed-up (Table 8.2). After these simulations, it is observed that the *des_round* function (the hot-spot function of DES) performs several accesses to data memory containing S-box. Each S-box contains 64 32-bit elements. These memory contents are known prior to the hot-spot execution. Existing studies [115] on exploiting such knowledge show that the runtime improvement can be stronger by including scratchpad memories within the CI. To experiment with such extensions, local scratchpad memory resource is appended to the rASIP description. A special instruction to transfer the S-boxes from data memory to the scratchpad memory is included. ISEGen is then configured to have up to 4 parallel scratchpad accesses. Each of these configurations produced different set of custom instructions. Since the scratchpad access is local to the re-configurable block, the interface constraints are not modified due to the access. The only modification required is for allowing the data transfer from the data memory to the scratchpad memory. The complete simulation and synthesis results for custom instructions with scratchpad access are given in Table 8.3.

Table 8.2   Simulation cycles: DES

Latency	Without custom instructions	With custom instructions	Speed-up
4	1563266	625306	2.5
Hidden	1563266	453397	3.5

Table 8.3 Custom instructions with scratchpad access: DES

Number of parallel scratchpad access	Speed-up	Minimum clock delay (ns)	Area (LUTs)
1	4.4	10.7	1342
2	4.2	9.72	1665
3	5.9	9.05	1638
4	6.0	9.05	1616

Considering no significant change in speed-up with other interface constraints (e.g. 3-input,1-output and 4-input,1-output), it is decided to keep the 2-input, 1-output interface setting for the rASIP fabrication. Since scratchpad memories improved the runtime performance considerably, it is also integrated. Number of parallel accesses can be increased or decreased post-fabrication depending on the available re-configurable block area. Note that the contents of the scratchpads can be physically implemented as flexible hardware tables, too – possibly resulting in higher speed-up in the data access.

Post-fabrication Design Space Exploration: In keeping trend with crypto-graphic applications, the hot-spot of Blowfish application does also contain accesses to pre-calculated memory elements. Those elements are loaded to the scratchpad memory. ISEGen identified various set of custom instructions for Blowfish. The interface restrictions from the pre-fabrication design as well as various scratchpad access configurations are fed to the ISEGen. The generated set of instructions are then appended to the rASIP description. The synthesis and simulation results (refer Table 8.4) demonstrate the prudence of pre-fabrication decisions.

For the GOST application, the hot-spot function is found to be relatively small, thereby providing little opportunity to speed-up. Even then, the improvement is visible without and with the scratchpad access. The results are summarized in Table 8.5. Interestingly, 2 parallel scratchpad accesses resulted in poor speed-up compared to 1 parallel access. It is observed that the ISEGen left some base processor instruction out due to GPR I/O restrictions (2-input, 1-output). The base processor instructions with a subsequent custom instruction incurred extra *nops* due to data dependency. This serves as an example of how the interface restriction can control the speed-up. By allowing increased number of scratchpad accesses a bigger data-flow graph could be accommodated in the CI, thereby avoiding the GPR restriction. Similar

Table 8.4 Simulation and synthesis results: Blowfish

Number of parallel scratchpad access	Speed-up	Minimum clock delay (ns)	Area (LUTs)
0	2.7	13.78	1456
1	3.3	8.72	1009
2	3.4	13.78	1221
3	3.5	8.43	1473
4	3.8	9.05	939

Table 8.5 Simulation and synthesis results: GOST

Number of parallel scratchpad access	Speed-up	Minimum clock delay (ns)	Area (LUTs)
0	1.02	10.45	1803
1	1.6	9.05	1554
2	1.5	8.23	1513
3	1.7	9.05	1575
4	1.8	8.43	1455

effect is visible for DES (Table 8.3), too. However, 4 scratchpad accesses masked the effect of sub-optimal GPR I/O decision.

The strong improvement in the application runtime in the post-fabrication phase shows the importance of flexibility, which could be offered by rASIP in comparison with the ASIP. Note that the custom instructions selected for the DES applications are different from the custom instructions selected for the Blowfish or GOST application, stressing the importance of post-fabrication flexibility. If the designer wanted to fabricate all the custom instructions together in a single ASIP, it would cover an additional area of 12,853 gates. Obviously, such instructions cannot be foreseen in the pre-fabrication phase either. The results also reflect that the improvement is strongly dependent on the application and a prudent selection of the pre-fabrication design constraints. Finally, the complete design space exploration, starting with a LISA description of the base processor, took few hours by a designer. This is a manifold increase in design productivity, while maintaining the genericity.

8.3 Experiments with VLIW-Based Architecture

In the case study with VLIW-based architecture, a commercial architecture targeted for multimedia applications is chosen. It serves two purposes. Firstly, it is a processor designed with a target group of applications in mind. Extending this processor to the rASIP class will give us an opportunity to verify the performance advantage that an rASIP can deliver. Secondly, the complexity of this processor makes it nearly impossible to explore the design space manually. This can show the advantage of having a good rASIP exploration framework.

TriMedia32 (TM32), a member of Philips TM1×00 processor series with a VLIW core, has been designed for use in low cost, high performance video telephony devices [192]. TM32 is based on a 5-slot VLIW architecture and a 6-stage pipelined execution to enable highly parallel computing in time-critical applications. TM32 issues one long instruction word each clock cycle. Each long instruction is comprised of five 48-bit instructions, each corresponding to one slot. The operations are similar to those of standard RISC processors. The operations can be optionally guarded. Some of the functional groups of operations supported by TM32 are 32-bit arithmetic, dual 16-bit and quad 8-bit multimedia arithmetic, floating point arithmetic, square root and division. Multiple operations of the same type can be issued in different slots. However, all operations are not supported in all slots. The processor is equipped with an array of 128 32-bit General Purpose Registers (GPRs). Except r0 which always returns the value of 0×0, and r1 which always returns the value of 0×1 corresponding to the boolean values of FALSE and TRUE, all registers can be freely used by the programmer.

For this case study, the complete TM32 is captured in the proposed rASIP description. From the description, the software tools, including the C-compiler are generated. The rASIP description of TM32 consists of 29366 lines of code. The automatically generated Verilog RTL description for pre-fabrication implementation is of 192524 lines.

The first group of applications, which is used for pre-fabrication rASIP exploration contains image processing kernels from various kinds of image processing algorithms:

- The image convolution (*img_conv*) kernel accepts 3 rows of x_dim input points and produces one output row of x_dim points using the input mask of 3 by 3.
- The image quantize (*img_quant*) kernel quantizes matrices by multiplying their contents with a second matrix that contains reciprocals of the quantization terms.
- The image correlation (*img_corr*) kernel performs a generalized correlation with a 1 by M tap filter.

The second group of applications, for post-fabrication rASIP enhancements are chosen from cryptographic application domain. One image processing kernel is also added in order to see the quality of pre-fabrication architectural decisions. The cryptographic kernels are Blowfish and GOST, the ones already utilized for case study with the RISC-based architecture. The image processing kernel is named *img_ycbcr*.

- This kernel (*img_ycbcr*) converts YCbCr (Y - luma (luminance); Cb - chroma blue; Cr - chroma red;) color space to RGB (Red - Green - Blue) color space, useful in many image and video applications.

Pre-fabrication Design Space Exploration: The position of the re-configurable block, its coupling with the base processor and the resource accessing schemes are developed during pre-fabrication design space exploration using the image processing kernels. The image convolution kernel revealed two hot-spots from application profiling. In the first one, a general purpose register is read and conditionally modified. In another hot-spot 6 parallel memory accesses are made, which cannot be supported by the architecture. The calculation followed by the memory accesses was already parallelized to a high degree by the VLIW slots. Therefore, the application runtime speed-up potential is low. The image quantization kernel showed a potential custom instruction, where two registers are read, multiplied, shifted and then written back to the destination register. The image correlation kernel suggested the requirement of a multiply-accumulate (MAC) operation. Further improvement is obtained by software pipelining the target loop manually, where the memory read of next iteration is combined with the current MAC operation. This required the re-configurable operations to be implemented in a branch of instruction tree, which does not affect parallel memory accesses.

From the pre-fabrication design space exploration, several conclusions about the final rASIP architecture are made. Foremost, the variety of custom instructions, even within one application domain, justified the use of a re-configurable unit with local decoding. This allows ample freedom to add new custom instructions. By using the coding leakage explorer, a branch in the coding tree with 44 free bits out of total 48-bit instruction is found. Out of these 44 bits, 7 bits of coding space is reserved for new instructions with up to 5 read-write indices for register access. The pre-fabrication application kernels justified the usage of up to 4 register read and 1 register write operation. The coding branch is also selected to be in such a place that, parallel memory accesses can be done. Finally, the frequent access of general purpose registers and effective usage of instruction-level parallelism in the overall architecture demanded the re-configurable unit to be tightly coupled with the processor. The availability of a large array of general purpose registers also ensures a low probability of resource conflict even though there is a tight coupling.

With the knowledge that the processor is going to be used for cryptographic domain of applications in post-fabrication mode, another extension is made to the architecture. Since the cryptographic applications are data-driven, a direct interface from the re-configurable unit to the data memory is established by putting a dummy memory access. This allowed the construction of complex custom instructions, including the memory access. Alternatively, local scratch-pad memories could be allocated in the re-configurable block [115]. This could be achieved by having dedicated array of internal storage elements in the re-configurable block.

The RTL implementation of the complete rASIP description is automatically generated. The base processor is synthesized with Synopsys Design Compiler [190] using 0.13 μm process technology of 1.2 V. The re-configurable block is synthesized with Synopsys FPGA Compiler [190] using Xilinx Virtex-II pro [191] (0.13 μm process technology of 1.5 V) as the target device. The pre-fabrication synthesis results are given in Table 8.6. The huge area of the base processor is mostly (78.28%) contributed by the 128-element 32-bit GPR array. The area of the re-configurable block is not constrained in the pre- or post-fabrication phase. The pre-fabrication simulation results are shown in Table 8.7. As expected, the application runtime speed-up is not very strong in case of the image convolution kernel.

Post-fabrication Design Space Exploration: During the post-fabrication enhancements, the interface constraints, which are fixed during the fabrication are fed to the ISEGen tool in order to obtain the custom instructions. ISEGen identified 3 custom instructions for the Blowfish encryption-decryption algorithm. The behav-

Table 8.6 Pre-fabrication synthesis results for TM32

Re-configurable block		Base processor	
Area	Minimum clock delay (ns)	Area (Gates)	Clock Delay (ns)
572 LUTs 12 multipliers	16.5	596790	4.6

Table 8.7 Pre-fabrication simulation results for TM32

Application	Cycles w/o CIs	Cycles with CIs	Speed-up
img_conv	76,838	66610	1.15
img_quant	4902	3622	1.35
img_corr	170,720	86822	1.97

ior of the custom instructions are embedded in the LISA model with the coding determined from the coding leakage explorer. The identified custom instructions are involved in several address calculation, memory accesses and arithmetic operations on the accessed data. The direct memory interfacing and the availability of internal storage elements allowed composition of large custom instructions with high speed-up. The flexibility in opcode space could be fully utilized for accessing the internal storage elements and the memory. For GOST algorithm, 5 custom instructions are determined, which satisfied pre-fabrication constraints. The custom instructions are further parallelized manually by the VLIW capabilities of the processor. The post-fabrication image processing kernel (img_ycbcr) revealed several hot-spots with conditional arithmetic and logical operations. It showed similar traits like previous image processing applications e.g. several parallel GPR read accesses, parallel memory access etc. The post-fabrication synthesis and simulation results are given in Tables 8.8 and 8.9 respectively. As can be observed, the runtime improvement is strong in the image processing kernel as well as in the cryptographic kernels. Relatively small improvement in GOST runtime is due to the its small-sized hot-spot function.

The speed-up in the application runtime for TM32 shows that defining optimum architecture for even a domain of applications is extremely difficult. There are improvements possible even for TM32, which is designed with multimedia applications in mind. Clearly, it is not possible to accommodate all the special-purpose in-

Table 8.8 Post-fabrication synthesis results for TM32

Application	Re-configurable block		
	Area	Minimum clock delay (ns)	Latency
Blowfish	1042 LUTs 254 registers	14.6	4
GOST	1133 LUTs 238 registers	12	3
img_ycbcr	312 LUTs 2 multipliers	16.5	4

Table 8.9 Post-fabrication simulation results for TM32

Application	Cycles w/o CIs	Cycles with CIs	Speed-up
Blowfish	728926	311646	2.34
GOST	182110	144982	1.26
img_ycbcr	12700	6316	2.01

structions together in one ASIP. rASIPs, by having a changeable instruction-set, can effectively address this issue. The experiments with VLIW processor also showed that the ISEGen is not capable of identifying custom instructions keeping the VLIW slots in mind. Parallelization of the custom instructions are done manually, resulting in higher speed-up. The other lesson learnt from this case study is that it is difficult to obtain high speed-up by using custom instructions for VLIW architectures since, one of the basic reasons of runtime improvement by custom instructions is their exploitation of parallelism in the application. To boost the runtime performance either sequential operators need to be grouped in a custom functional unit or further parallelism need to be explored or both. To establish opportunities for further parallelism, the re-configurable block must be equipped with high number of data-transfer interfaces. Actually, the less number of data-transfer resources did not allow to model a particular custom instruction of GOST application, as explained in the following.

The portion of GOST application that was subjected to optimization can be found in Fig. 8.1. There are two code sections – the code in the left part represents the candidates for implementation as custom instruction and the right part represents their corresponding LISA implementation. For clarity, the DFG representation of candidate 1 is also shown. The candidate 1 is separated into three phases: the address

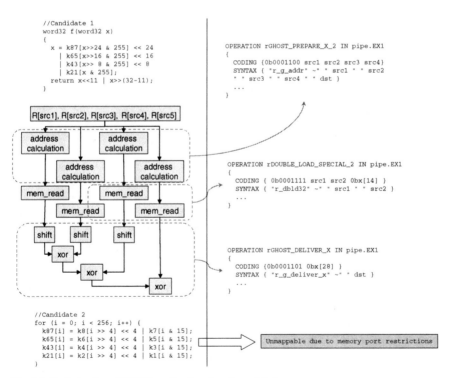

Fig. 8.1 Optimization of GOST application: interface violation

generation, the memory access and the data processing. The candidate 2 contains multiple lines of code where two memory reads and one memory store is needed. Having three memory accesses, even with special instructions it is not possible because of only two available memory ports per clock cycle from the re-configurable block. Unlike the RISC case study, no scratchpads are allocated locally to the re-configurable block. Therefore, candidate 2 cannot be implemented with a chance of gain.

Even for the proposed division of custom instructions in Fig. 8.1, the first custom instruction requires 5 simultaneous register read paths, while the interface can accommodate only 4. This means, that there is a lack of the extra 32-bit data bits, and the enable/activation bits for that extra data path. Therefore, given the pre-fabrication interface decision, the only option is to split up the custom instruction for address generation into two smaller instructions. This is what is done in the case study and the results are as shown in the Table 8.9. This simple trade-offs establish the importance of prudent pre-fabrication design choices aided by a sound understanding of target post-fabrication application domain/evolution.

8.4 Experiments with Coarse-Grained FPGA Exploration

The four algorithm kernels selected for this experiment are FFT Butterfly (BFLY), FIR (8-tap), DES and IDCT. FFT and FIR are well-known algorithm kernels widely used in communication and digital signal processing. DES is a block cipher algorithm used in the previous case studies. The block targeted for FPGA exploration is the computation-intensive part of the DES namely, DES-round. IDCT is a fourier-related transformation, often used for signal and image processing applications, especially for lossy data compression. It has two components reflecting similar traits, namely IDCT-row operations and IDCT-column operations. For the experimentation described in the following, IDCT-row function is taken.

Delay Model: In the entire case study section, a cycle-based cost model with inter-cluster routing delay set to 2 cycles and intra-cluster routing delay set to 1 cycle is used (DM1 cost model in [181]). Only in the case of heterogeneous FPGAs, where *pseudo cluster boundaries* are set up, the inter-cluster routing delay is set to be same as the intra-cluster routing delay, both being 1 cycle.

Attempts of Modelling Existing Architectures: In order to compare between different architectural styles, several features of well-known coarse-grained re-configurable architectures are first modelled. The architectural features are summarized in the Table 8.10. For FPGA-1, the cluster-level connectivity of MESH-1 is used, whereas for the rest the cluster-level connectivity is not relevant. The basic element used in all these architectures is a 32-bit ALU with arithmetic and logical operators inside those. The input and output ports of the basic elements can be registered or bypassed. The connectivity style and connectivity strides are also indicated in the table. For example, the MATRIX architecture supports a connectivity style of nearest neighbour (NN) with a stride of 1, mesh with a stride of 2, row-wise

Table 8.10 Instances reflecting known coarse-grained FPGAs

Architecture	Cluster Size	FPGA Size	FPGA-level Connectivity	Reflecting Topology of
FPGA-1	2×2	8×8	MESH-1	DReAM
FPGA-2	1×1	8×8	NN-1, MESH-2, ROW-4, COL-4	MATRIX
FPGA-3	1×1	8×8	MESH-1, ROW-1, COL-1	MorphoSys

(ROW) and column-wise (COL) both with a stride of 4. Actually, the row-wise and column-wise 4-hop connection in the MATRIX architecture is present in alternative fashion, which is simplified for this study. Note that, several of the presented architectures support multiple styles and strides together at one hierarchical level. Instead of building a template library of all possible styles, the approach proposed in this work is to break up the complex routing styles into group of overlapping ones.

Two of the algorithm kernels are synthesized with the aforementioned FPGAs. From the results given in Table 8.11, it is not hard to find out the following notes. Firstly, in the architecture which is close to DReAM [193], for both applications, the number of clusters used after placement and routing is much more than the number of clusters used before that. The reason for this is, since MESH connectivity is used in FPGA level, a lot of extra clusters are used for routing purpose. How many extra clusters are used for routing depends on the kind of connectivity style and the routing capacity of clusters. Therefore, further exploration of connectivity style in FPGA level is necessary to achieve better results. Secondly, in both the architectures FPGA-2 and FPGA-3, a cluster size of 1×1 is used. In such a case, the in-cluster configuration does not call for any exploration. The performance after placement and routing depends on the FPGA-level connectivity. A better performance is achieved for the architecture with MATRIX-like connectivity due to more availability of routing resources.

Effect of Connectivity: For the FPGA to have a right balance of performance and flexibility, it is imperative to select the basic elements, routing architecture and structural topology prudently. With different applications, it turned out that the performance varies with different architectures. To understand this variation, the exploration started with a simple architecture with a cluster size of 2×2. In each cluster, three basic elements are used namely, ALU (for arithmetic operations), CLB (for logic operations) and MULT (for multiplication). The cluster is optionally equipped with a multiplexer block to enable control flow mapping. The arrangement of the

Table 8.11 Synthesis results with the topology of known coarse-grained FPGAs

Architecture	Application	Number of Clusters before P&R	Number of Clusters after P&R	Critical Path (cycles)
FPGA-1	IDCT-row	26	42	53
	DES-round	12	23	23
FPGA-2	IDCT-row	58	62	24
	DES-round	28	31	24
FPGA-3	IDCT-row	58	67	32
	DES-round	28	35	30

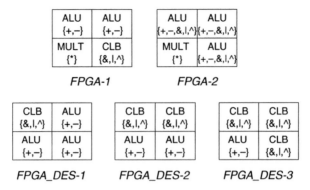

Fig. 8.2 FPGAs for exploration

elements inside cluster is as shown in the Fig. 8.2 (FPGA-1). On that basis, the FPGA-level connectivity is varied to obtain the results as presented in Fig. 8.3 and in Fig. 8.4. Clearly a rich interconnect structure reduces the critical path but, not for all kernels. Interestingly, for DES-round, a connectivity style of NN-1 achieves better critical path as well as cluster count than a connectivity style of {MESH-1, NN-2}. This clearly reflects the application data-flow organization, which is much denser than can be supported by MESH with 2-stride NN.

Effect of Functionality of Element: In this experiment the functionality of the elements are altered. This can be performed easily by modifying the OPERATOR_

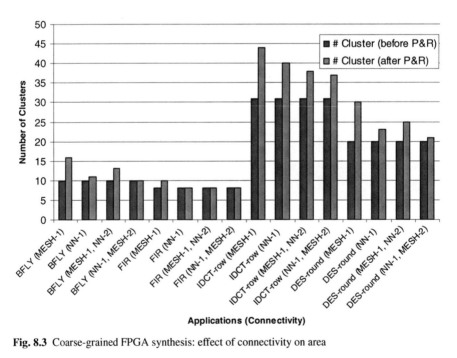

Fig. 8.3 Coarse-grained FPGA synthesis: effect of connectivity on area

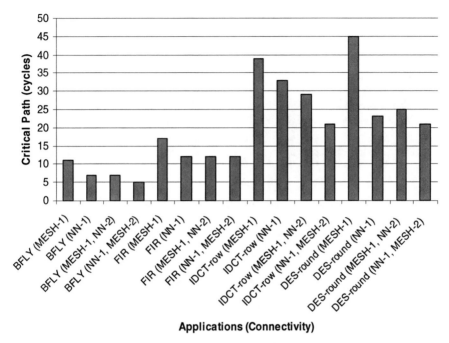

Fig. 8.4 Coarse-grained FPGA synthesis: effect of connectivity on critical path

LIST of an element. Here, the operators defined in CLB are moved into ALU and the CLB is replaced with another ALU. Now, the architecture becomes the FPGA-2 of Fig. 8.2. The results are shown in the Fig. 8.5.

Compared to the original structure, the functionality modification allowed the arithmetic operations and logic operations in the application to fit into the same element. Because one more ALU element is now available inside the cluster, chances of more arithmetic or logic operations to be put into one cluster is increased. Therefore, better mapping results are easily found in BFLY, IDCT-row and DES-round. However, there is not much difference for FIR in terms of the critical path. This is since there are no logical operations in FIR. In terms of number of clusters, FIR is checked to have exactly the same results for both FPGA-1 and FPGA-2.

Effect of Varying Number of Elements in Cluster: To show the effect of varying element numbers in a cluster, the DES-round kernel is chosen. In DES-round, there are only logic and arithmetic operations, which are distributed in a ratio of roughly 3 to 1. In this experiment, three architectures from Fig. 8.2, with all of those having {NN-1,MESH-2} connectivity at FPGA level. From the results (refer Table 8.12), it can be observed that, a good architectural decision is based on the proper application characterization. The selection of element type and number of elements inside cluster should follow the basic characteristics of application e.g. the ratio of operators inside application. Here, when the ratio of elements of corresponding type is close to the ratio of operators in application, a better synthesis result is achieved.

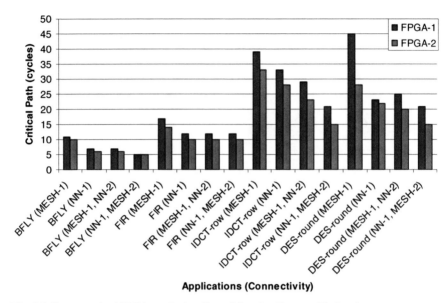

Fig. 8.5 Coarse-grained FPGA synthesis: effect of functionality on critical path

Area-optimized CG-SMAC: By trading-off between the results of all the kernels and taking the effects which are analyzed above, FPGA-2 is found to be the best performing one. For area consideration, the MULT is kept as a separate element out of ALU. Considering the characteristics of BFLY and FIR application, only one MULT is arranged inside cluster. Since the logical and arithmetic (w/o multiplication) operations dominate in BFLY, IDCT-row and DES-round, three ALU which includes both logic and arithmetic operations are put inside cluster. An NN-1 connectivity in cluster level and an {NN-1, MESH-2} connectivity at FPGA level is used. For this FPGA, area-optimized version of CG-SMAC is applied to observe the effect. The results are recorded in Table 8.13. The synthesis results without area optimization are indicated within square brackets. Better area results are obtained in all cases. However, a degradation of critical path is also observed as expected. This area-optimized version of CG-SMAC can be employed suitably when the FPGA size is fixed beforehand and/or when the delay constraints are less strict. An interesting follow-up work can be to perform area-optimization in non-critical paths as in [68].

Table 8.12 FPGA synthesis: effect of diversity

Architecture	Number of Clusters before P&R	Number of Clusters after P&R	Critical Path (cycles)
FPGA_DES-1	20	22	23
FPGA_DES-2	14	14	16
FPGA_DES-3	11	11	13

Table 8.13 FPGA synthesis: area-optimization

Application	Number of Clusters before P&R [w/o optimization]	Number of Clusters after P&R [w/o optimization]	Critical Path (cycles) [w/o optimization]
BFLY	8 [9]	8 [9]	7 [5]
FIR	7 [8]	7 [8]	11 [10]
IDCT-row	23 [26]	28 [31]	21 [15]
DES-round	12 [14]	12 [14]	23 [15]

Heterogenous FPGA Synthesis: For experimenting with heterogenous FPGA structures, the application FIR is chosen with the architectures taken from Fig. 8.6. An overall MESH-1 connectivity style is chosen with the inter-cluster and intra-cluster routing delay set as 1 cycle. For the H-FPGA-2 architecture, the MULT elements are placed more sparsely, which made the routing path longer. This resulted in higher number of clusters as well as longer critical path (Table 8.14).

Runtime Complexity: Though the algorithms used in the FPGA synthesis flow are computation-intensive, the relatively less complexity of interconnects in FPGA compared to fine-grained FPGAs allowed all the presented case studies to be synthesized in reasonable time. In a AMD Athlon Dual Core Processor (each running at 2.6 GHz), the case study applications finished within 1 (FIR, 15 operators) to 15 min (IDCT-ROW, 58 operators) for FPGA-2.

FPGA Implementation: The architectures used in this case study are synthesized to obtain RTL description, followed by gate-level synthesis with Synopsys Design Compiler [139]. For comparison's sake, the synthesis results for the architecture FPGA-2 are presented here. FPGA-2 is synthesized with total 25 (5×5) clusters. For the designer-specified connectivity, total 2542 configuration bits are required to control FPGA-2. After gate-level synthesis, the entire architecture met a clock constraint of 5 ns (with register attributes at element's outports) for

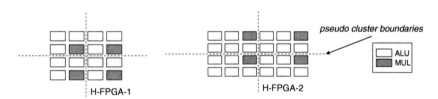

Fig. 8.6 Heterogeneous FPGAs

Table 8.14 Heterogeneous FPGA synthesis

Architecture	Number of Clusters after P&R	Critical Path (cycles)
H-FPGA-1	20	8
H-FPGA-2	22	10

130 nm process technology (1.2 V) and occupied an area of approximately 3.77 mm^2 of which 1.62 mm^2 area is consumed by the FPGA-level routing alone. It should be noted that this synthesis figures are bound to improve significantly after physical optimization and by using special library cells, evidently for, the routing architecture.

8.5 rASIP Modelling for WCDMA

To aggressively support wide-ranging and continuously evolving data communication standards, future wireless receivers are predicted to be cognizant of its operating environment. A major factor enabling this intelligent receiver is its flexibility during execution. While processors support flexibility via ISA, the fast-changing communication algorithms may require excellent performance from a processor throughout its evolution. Furthermore, the processor needs to deliver performance across varying standards, which is not an easy task. Generally, such wireless platforms are referred as Software-Defined Radios (SDRs). rASIPs appear to be an important design candidate for being used in such flexible and adaptive wireless receivers. Consequently, researchers are proposing novel architectural solutions to combine various wireless receiver algorithms under a single hood [194], utilizing modern rASIPs as IP blocks to deliver SDR solutions [195] and even proposing fully re-configurable solutions [196] to address the challenge of performance and flexibility raised by SDR. For the final case study, a representative wireless protocol is chosen. Wideband Code Division Multiple Access (WCDMA) [197] is one of the most common 3rd generation cellular protocols, which an SDR solution must support. In the remaining part of this section, it is shown how WCDMA is analyzed and steadily converged to a solution via several iterations – all along using the rASIP tool-flow proposed in this book.

For this case study, a high-level language (C) implementation of UMTS WCDMA receiver is obtained from CoWare Signal Processing Designer [198]. The receiver's algorithmic block diagram is shown in the Fig. 8.7. The upper part of the figure shows a generic algorithmic flow, out of which from the *match filter* till the *maximum ratio combiner* is selected as the target application. In the lower part of the figure, the implementation of each rake finger in the WCDMA receiver is shown in detail. From the C implementation, various functional blocks are identified and grouped together to represent the algorithmic blocks. As can be observed, 4 rake fingers are deployed for the receiver. The incoming chip rate is 3.84 Million chips per second with 15 slots per frame and each slot carrying 2560 chips. The channel oversampling factor is 4 and the spreading factor is 64. The *match filter* employed here, is implemented using radix-2, 128-point FFT adjusted by overlap-add method. The complexity of the standard time-domain convolution (64-tap FIR filtering) justifies using a frequency-domain multiplication instead. Surely, by including effective memory organizations and special instructions, the frequency-domain operation can be further boosted [199].

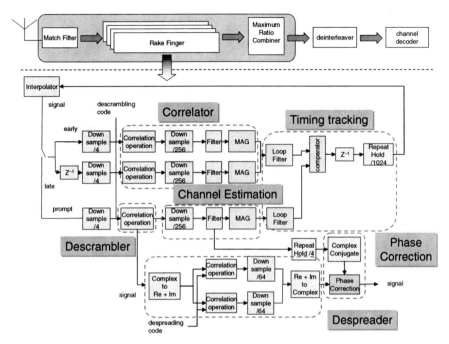

Fig. 8.7 Algorithmic block diagram: UMTS WCDMA receiver

8.5.1 Profiling

Apart from algorithmic study of the implementation, a direct analysis of the computational requirements of the WCDMA implementation is undertaken. This is performed by subjecting the C implementation to the architecture-independent profiling tool [105]. From the profiling results, a coarse estimation of the number of operations per algorithmic block is obtained. From the algorithmic analysis, it is also simple to determine the sampling rate for each block. By putting these two pieces of information together, a log-log graph is obtained (refer Fig. 8.8). The figure clearly shows that the computational complexity requirements of various blocks are different, sometimes by wide margins. Considering the base processor performing in the range of few hundred MOPS, the strategy to meet the timing deadline is as presented in the figure. For low-complexity operations, direct software implementations suffice. For high-complexity blocks, as in the match filter, special-purpose custom instructions must be devised along with hardware extensions. For some complex blocks, the ISA extensions may not be sufficient in the long run. This is since, the algorithm itself may change. For such cases, coarse-grained building blocks are created, which can be flexibly connected to each other. This is typical of the correlator block, for which coarse-grained FPGA implementation can be chosen. Another decision to be made from the profiling is what kind of basic architecture is to be used.

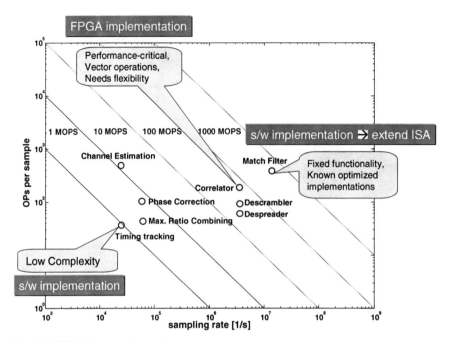

Fig. 8.8 WCDMA complexity analysis

For embedded processors, sophisticated hardwired scheduling is unusual, leading to simple-scalar processors. Nevertheless, to support parallelism inherent in many algorithmic blocks, SIMD processing is required. By analyzing the complex blocks of WCDMA (e.g. match filter, correlator) in detail, it is found that high-degree of coarse parallelism do exist in this blocks. For example, correlation is performed in parallel for 3 incoming sample streams (early, prompt, late) for 4 rake fingers. On the other hand, several scalar computations are also embedded within this parallelism. Transferring streaming data among different blocks, scaling the data, buffering the data are some examples of such computations. A VLIW processor would require high number of independent buffers and their synchronization in order to reap the benefit of the algorithm-level parallelism, resulting into high area consumption. Furthermore, it would be difficult to determine fine-grained ILP to accommodate the scalar computations. On the basis of this, a RISC processor with several vector computing units is proposed as a starting point for the architecture exploration of WCDMA. Depending on the exploration outcome, a fully VLIW architecture may be designed.

8.5.2 Base Architecture

The base architecture for this case study is termed as IRISC. IRISC is connected with two synchronous memories namely, program memory and data memory. The pipeline of IRISC is distributed across 5 stages, *prefetch, fetch, decode, execute* and *writeback*. The architecture supports bypass and interlocking mechanism to handle data hazards. The IRISC instruction set consists of conditional and non-conditional instructions. Apart from regular load-store instructions, IRISC supports arithmetic instructions like addition, subtraction, multiplication; boolean instructions like and, or, xor, right shift, left shift; comparison instructions and loading and storing of complex data-type. IRISC architecture contains 16 32-bit general purpose registers and no special-purpose register.

8.5.3 Hardware Extensions

For supporting the frequency-domain multiplication operations in the match filter, several hardware extensions are naturally done to the IRISC. It is argued that the match filtering is mostly a stable algorithmic block, where limited flexibility (e.g. number of taps in the filtering) would be sufficient. For supporting that, a dedicated coefficient memory is introduced. Special registers for holding the intermediate results in the Fast Fourier Transformation (FFT) are introduced, too. The basic data memory is also extended with a dual port one, to continuously feed the functional units operating in parallel. To reduce the amount of port access from memory, the real part and the imaginary part of the incoming complex data are stored together in the same memory location, where the upper half represented real part and the lower half represented imaginary part. Recognizing a huge number of loop operations, a dedicated set of registers are included for supporting zero-overhead-loop mechanism. Further special purpose registers are introduced to support parallelism in the rake fingers, which store the result of multiply-accumulate operations. These registers could have been put simply into the pipeline registers, too. The advantage of dedicated registers is that non-streaming data can be suitably stored and retrieved from these. In order to support custom instructions without increasing the critical path, 4 new pipeline stages are introduced. Notably, the additional pipeline stages are used by few special purpose instructions only. The regular instructions from basic IRISC still finishes by the 5 pipeline stages. The additional special instructions are allowed the extra stages as those involve multi-cycle memory-based execution. By increasing the number of stages, more latency (and thereby, more special instructions) is allowed without hampering the previously achievable throughput.

8.5.4 Instruction Set Architecture Extensions

While moving from a basic RISC architecture to a specialized ASIP, IRISC incorporated several extensions in its instruction set. In the following, a summary of these extensions and its target algorithmic block in WCDMA are given.

Memory-Memory Instruction For the complex filtering operation, multiplication with the coefficients is an integral part. As the coefficients are stored in dedicated memories to speed-up the parallel execution, special memory-memory instruction for multiplication is also designed.

Zero-Overhead Loop This mechanism is utilized over the complete WCDMA. A special instruction is designed to load the number of iterations and the number of subsequent instructions, which are to be repeated.

Complex Butterfly Instruction For speeding up the match filter block, frequency-domain multiplication is used. This necessitates transformation from time-domain to frequency-domain and the reverse. This is performed via FFT, of which butterfly operation is a key component. A special instruction for performing complex butterfly is designed. The instructions' datapath and how it is distributed over the pipeline stages are shown in the Fig. 8.9. For storing the intermediate results of FFT operations, a set of dedicated cache registers (32 32-bit unsigned registers) are used. A special purpose instruction (repeated using zero-overhead loop) dumps this cache registers into memory.

Complex Arithmetic Operator All over the WCDMA, complex arithmetic operations are rampantly used. Be it the multiplication in descrambler or the addition in the correlator, it is simply advantageous to have complex

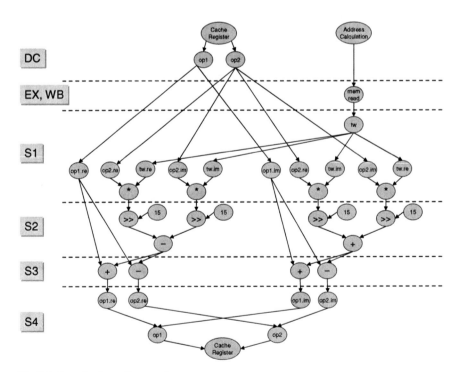

Fig. 8.9 Complex butterfly operation

addition, multiplication, multiply-accumulate, magnitude calculation, conjugation operators. Obviously, different instructions require different latency and those are distributed across pipeline stages accordingly. The data movement instructions transfer data from special-purpose registers to memories and vice-versa aiding the complex arithmetic operators. In addition to the complex instructions, modulo instructions are also supported for enabling the down-sampling operations.

LFSR-based Code Generation The descrambling code in the WCDMA is generated via a pseudo-random sequence generation process, employing Linear Feedback Shift Register (LFSR). A software implementation of LFSR is surely inefficient due to its fine-grained hardware design. Furthermore, the random sequence of code may change from implementation to implementation, demanding a certain flexibility. A custom instruction for generating the descrambling code is designed. It comes with two arguments namely, the initial seed and the generator polynomial – both of which can be modified by the software developer.

Bit Reversal This is not employed as a stand-alone instruction but, is tagged with several data movement instructions. Bit reversal is particularly needed for the butterfly operation of FFT (used in the match filter).

Data Movement Various special-purpose registers are used for keeping the supply of data to the functional units of IRISC. Several special instructions are designed to move the data across these storage locations. To allow blockwise data transfer, automatic address increment after each data transfer or bit-reversal during the data transfer is also supported. In the case of last stage of forward FFT operation, the data is moved to memory and on-the-fly multiplication with the coefficient memory is performed.

8.5.5 Software Extensions

The aforementioned extensions naturally call for an upgrading of the software implementation to fully exploit the additional hardware and the special-purpose instructions. The organization of the loops and functions in the reference software implementation are alternated to utilize the custom instructions to the full extent. The complex and demanding blocks of the algorithm e.g. match filter, despreader, descrambler and correlator are particularly modified with assembly-level coding. This includes the forward and inverse FFT operations. The instructions, which

Table 8.15 Synthesis results for IRISC

Architecture	Minimum Clock Delay (ns)	Area (KGates)	WCDMA Execution Time (cycles)	Energy (μJ)
IRISC-Basic	2.0	27.46	126988124	3936.6318
IRISC-ASIP	2.0	79.76	8548035	521.4301

require several cycles to be completed, need to be buffered with *nops* following them. This is also achieved by effectively utilizing the inline assembly feature of LISA compiler.

8.5.6 Synthesis Results

After the design exploration phase, the generated RTL description is synthesized with Synopsys Design Compiler [139] using 90 nm process technology (refer Table 8.16). The synthesis results of IRISC processor before and after the extensions reveal significant gain in application runtime performance (simulated for 1 slot) and thereby the energy figures. Expectedly, area consumption is much higher. By introducing additional pipeline stages and careful distribution of operations, the frequency of execution remained at the same level as in the original.

Table 8.16 Synthesis results for IRISC with SIMD_CMAC

Architecture	Minimum clock delay (ns)	Area (KGates)	WCDMA execution time (cycles)	Energy (µJ)
IRISC-ASIP	2.0	79.76	8548035	521.4301
IRISC-ASIP (with SIMD_CMAC)	2.0	132.87	7160311	545.6157
IRISC-ASIP (with clock gating)	2.0	131.46	7160311	519.8386

Among this overall runtime speed-up, it is interesting to look at the critical blocks of the algorithm. The match filter portion of WCDMA uses FFT operation. The FFT used to take 265,381 cycles in the original IRISC processor. Following the extensions, it executes in merely 1047 cycles, which is an improvement of runtime by 2 orders of magnitude.

8.5.7 Extending Towards rASIP

From the algorithmic analysis and profiling results, the correlator block is found to be most suitable for a coarse-grained FPGA implementation. Correlation blocks need to be quick in execution as those are part of the loops responsible for timing tracking and phase correction. The loops impose hard deadlines on the timing. To meet these deadlines, a straightforward solution is to have dedicated ASIC implementation of correlator with the control blocks residing in the base processor. The major issue with the ASIC implementation is that it cannot serve as a building block of the system even if a minor change in the channel size or configuration is altered. Correlators are used for overcoming the problem with multi-path propagation in WCDMA. Researchers have proposed linear combination of early and late correlation with flexible delay [200] to address this problem more accurately. Future high-sensitivity receivers may require further changes in the correlator channel, as indicated in [201]. Clearly, the correlator implementation needs to be highly flexible and also efficient. Coarse-grained FPGA offers the perfect choice for this.

There are few issues to be resolved here. First, which parts of the algorithm to be moved to the coarse-grained FPGA and which parts not. Second, how to distribute the control and datapath of the re-configurable instructions. Finally, which structure of the coarse-grained FPGA fits the correlator block best.

The correlator block is present for three signals (early, prompt, late) for all the 4 rake fingers. Its task includes a complex MAC operation (denoted correlation operation in 8.7), down-sampling, filtering and magnitude calculation. For correlator block, the complex MAC operation is of less complexity as the multiplication is done with +1 or −1. The complexity is driven by the filtering operation, in this case which is a 6-tap FIR filter. To run the filtering operations in parallel, various groupings (out of the 12 parallel filters) are possible. To keep the area within limits and at the same time driving performance higher, 3 parallel functional units (for early, prompt and late signal) are conceived within the coarse-grained FPGA. Each functional unit represents a particular signal, for which 4 parallel filtering operations are performed, corresponding to each rake finger. The functional units are triggered by a special instruction, termed *SIMD_CMAC*. The instruction basically consists of a complex MAC operation, which is repeated 6 times using a zero-overhead loop mechanism. The instruction reads and writes from/to dedicated accumulation registers, which the base processor needs to update. New custom instructions are designed, as before, to facilitate the data transfer with these special registers.

Before proceeding with the coarse-grained FPGA modelling, it is interesting to look into the potential speed-up of the special instruction. After augmenting the IRISC-ASIP with SIMD_CMAC instruction (without moving it to re-configurable block), the simulation and synthesis steps are performed. The achievable frequency, energy, area figures provide an estimation of the performance if the IRISC would be fabricated as a fixed processor. Clearly, the area increment is heavy – accounting for the additional vector complex MAC units and special-purpose registers. However, it reaps benefit in the number of simulation cycles, particularly for the target correlator block – where a speed-up of 7.5 times is achieved. The increase in runtime performance is not reflected in the energy consumption, primarily due to the increased power consumption. This is reasonable due to the addition of more functional units and several special-purpose registers, which consume considerable power in the idle states. An aggressive power minimization approach with operand isolation and clock gating is required to check that overhead. For validating this point, a latch-based clock-gating is applied to all the special-purpose registers, which are added solely to facilitate the SIMD_CMAC instruction. The resultant energy figures show better results compared to the original ASIP. Surely, further benefits can be achieved by performing operand isolation over the special functional unit and clock-gating the dedicated pipeline registers. Interestingly, the area figures reduced slightly – likely due to the replacing of multiplexer-based enable signals with much smaller AND gates.

8.5.8 Coarse-Grained FPGA Exploration for SIMD_CMAC

The datapath of SIMD_CMAC instruction consists of four parallel slots, one of which is presented in the Fig. 8.11. It is built of boolean and arithmetic operators, with almost similar proportion of each. The datapath presented in the Fig. 8.11 represents one complex MAC operation. With this structure in mind, various cluster contents are chosen (presented in the Fig. 8.10) and corresponding connectivity is varied. With a delay model of 1-cycle intra-cluster delay and 2-cycle inter-cluster delay, the critical path and the area results are presented in the Table 8.17.

The cluster contents for FPGA_CDMA-1 and FPGA_CDMA-2 are chosen to reflect the proportion of various operators in the target datapath. Both of these fared quite poorly. The post placement-and-routing cluster count increased steeply and the critical path's visualization revealed that the routing network is complex, even with a rich connectivity style in the FPGA level. This can be explained from the algorithm's operator distribution. Though the combinational and arithmetic operators are equally proportioned in the datapath, those are located sparsely. Therefore, a single cluster with equally divided operators does not give any benefit. By merging the operators into a single cluster (FPGA_CDMA-3, FPGA_CDMA-4), both the critical path and the number of clusters are reduced. With increasing connectivity density, lower critical path and cluster is almost always achieved. For the complete SIMD_CMAC, the bitstream configuration is generated only for one slot and repeated across the structure.

After performing placement and routing of the target datapath, the arrangement of clusters for FPGA_CDMA-3 using two different connectivity styles are shown in the Fig. 8.12. It can be observed that the mesh type of connectivity required several clusters to communicate using empty clusters, thus creating a longer critical path. The nearest neighbour connectivity style, on the other hand, provides a more compact placement organization.

Once the complete datapath is fit into the given FPGA dimensions, the communication between clusters and/or within clusters can be suitably timed by setting/resetting the configuration bit controlling bypass/register mode at the output ports. Thus, a complete synthesis of the FPGA can be performed to get an approximation of the total length of the critical path and then it can be pipelined at suitable places. A possible pipelining scheme is indicated in the figure using red-colored

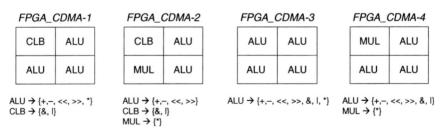

Fig. 8.10 Cluster structures for coarse-grained FPGA exploration

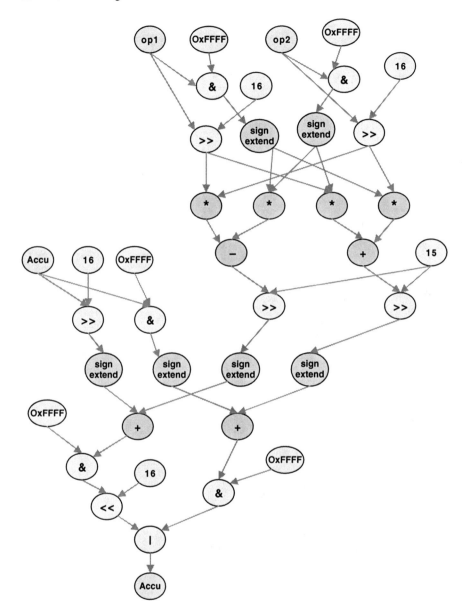

Fig. 8.11 Datapath of SIMD_CMAC instruction

dots. In case of the FPGA_CDMA-3 architecture, a gate-level synthesis of the generated RTL description is performed with Synopsys Design Compiler [139] using 90 nm process technology. This is done to estimate the performance of a single cluster. Routing architecture will not give any reliable estimate of the actual pass-transistor structure it should be made of. The complete cluster consumed an area of 21.02 KGates and met a timing constraint of 4.0 ns.

Table 8.17 Coarse-grained FPGA exploration

FPGA	Number of clusters before P&R	Number of clusters after P&R	Critical path (cycles)	Connectivity style (FPGA-level)
FPGA_CDMA-1	41	73	37	MESH-1
	41	62	29	NN-1
	41	62	25	MESH-1, NN-2
	41	52	25	NN-1, MESH-2
FPGA_CDMA-2	42	74	39	MESH-1
	42	68	31	NN-1
	42	61	21	MESH-1, NN-2
	42	53	21	NN-1, MESH-2
FPGA_CDMA-3	34	57	31	MESH-1
	34	49	23	NN-1
	34	49	19	MESH-1, NN-2
	34	43	17	NN-1, MESH-2
FPGA_CDMA-4	38	59	35	MESH-1
	38	53	27	NN-1
	38	57	21	MESH-1, NN-2
	38	45	17	NN-1, MESH-2

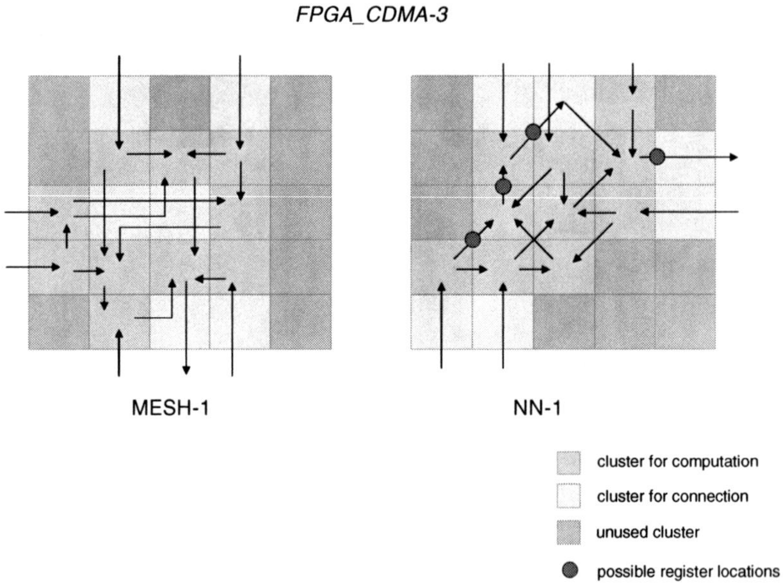

FPGA_CDMA-3

MESH-1   NN-1

cluster for computation
cluster for connection
unused cluster
● possible register locations

Fig. 8.12 Post place-and-route arrangement of SIMD_CMAC

8.6 Study on Design Space Exploration Efficiency

In this section, the focus of the case study is on demonstrating the *design efficiency* of the proposed tool flow, i.e. how different design alternatives can be quickly explored using the proposed methodology. For these experiments, the fixed base-processor is a simple RISC processor, named LTRISC, with a 5 stage pipeline and 16 GPRs. This processor is extended with a FPGA fabric for architecture exploration.

At the beginning of the case study, six different application kernels were selected for analysis. These kernels were selected from two prominent multimedia software suites: *mediabench II benchmark* and *X.264 codec implementation for the H.264 standard.* The selected kernels were: Inverse Discrete Cosine Transform (IDCT) from MPEG2 decoder, Inverse and Forward Discrete Cosine Transform (IDCT/DCT) from JPEG software, Sum of Absolute Differences (SAD) from H.263 and H.264 encoder, and Sum of 8×8 Hadamard Transformed Differences (SHTD) from H.264 encoder. The DCT/IDCT and SAD are chosen because various implementations of these kernels are embedded in a large number of media applications. SHTD was selected because it is one of the most computation intensive parts of the H.264 encoder.

The exploration flow mainly consisted of three steps – *ISE identification, FPGA exploration* and *interface exploration.* In the *ISE identification* phase, each application kernel was partitioned into a set of ISEs which were identified using a mixture of manual algorithm analysis, profiling with μProfiler [105] and automatic ISE identification. In the FPGA and interface exploration phases, the identified ISEs were inserted into the application code and simulated on the rASIP ISS to obtain speed-up results.

The FPGA exploration consisted of synthesizing the identified ISEs to various FPGA fabrics with different logic elements, topologies and connectivities. This step was used to characterize the different ISE sets in terms of cluster usage and critical paths. At the same time, this step provided hints on the best FPGA fabric for a given application kernel. The interface exploration consisted of bench-marking a variety of interfaces, as well as FPGA internal storage structures. In the MPEG-2 IDCT case, these two explorations were carried out independently, i.e. the FPGA was kept fixed during the interface exploration phase. In the other cases, the interface exploration was carried out only for the FPGA and ISEs short-listed through the FPGA exploration. Finally, the results of these two explorations were combined together to determine the best FPGA structure, interfacing options and ISEs for a given application kernel.

The next subsections present the different design points that are explored. For the MPEG2 IDCT routine, the FPGA and interface exploration are described in detail. For the other benchmark kernels, the configurations which achieve the best performance are presented. During the FPGA exploration, a cycle-based cost model with inter-cluster routing delay set to 2 FPGA clock cycles and intra-cluster routing delay set to 1 FPGA clock cycle was used (DM1 cost model in [181]). The gate-level synthesis results for LTRISC and FPGA clusters were obtained using a 130nm technology library and Synopsys Design Compiler. For all different design points, the base processor met a clock constraint of 2.5 ns, and the FPGA met a clock constraint of 5 ns, i.e. the base processor clock was running two times faster than the FPGA clock. All the critical path results in the next sections are reported in terms of base processor clock cycles.

8.6.1 MPEG2 IDCT Kernel

The IDCT routine from the MPEG2 decoder has two loops which process a 8×8 element data-array in row-wise and column-wise fashions. These loops have extremely similar structures and can be covered by the same set of ISEs. Two sets of ISEs were identified for these loop kernels after careful analysis. The first set contained 4 ISEs (named ISE1 through ISE4), while the second set had only one ISE which encompassed the entire DFG of the kernels.

FPGA Exploration: Algorithm analysis and μprofiling showed that the loop kernels use various arithmetic operations – mostly additions and subtractions, and some multiplications (around 15% of the total operators). Therefore, only ALUs and Multipliers were selected as PEs in the FPGA fabric. Three different cluster topologies (IDCT-1, IDCT-2 and IDCT-3 in Fig. 8.14) with 2×2, 2×3 and 3×3 PEs were explored. Nearest Neighbor (NN) connectivity scheme is used inside a single cluster, and explored NN, {Mesh-1, NN-2}, and {NN-1, Mesh-2} configurations for inter-cluster communication. The critical paths and cluster usage for the two ISE sets with different FPGA configurations are shown in Table 8.18.

Interface Exploration: For MPEG-2 IDCT, interface exploration was carried out with all 9 FPGA configurations listed in Table 8.18. For the first ISE set, the following set of interface options for each FPGA configuration are tried:

Table 8.18 FPGA exploration for MPEG2 IDCT

FPGA configuration		Partition I					Partition II(full DFG)	
Cluster name	Connectivity style	Critical path (cycles)				Number of clusters	Critical path (cycles)	Number of clusters
		ISE1	ISE2	ISE3	ISE4			
IDCT-1	NN-1	12	4	1	1	29	27	34
	Mesh-1, NN-2	11	4	1	1	28	23	33
	NN-1, Mesh-2	10	3	1	1	28	18	30
IDCT-2	NN-1	11	3	1	1	28	17	32
	Mesh-1, NN-2	9	4	1	1	27	17	30
	NN-1, Mesh-2	8	3	1	1	27	14	27
IDCT-3	NN-1	11	3	1	1	31	19	34
	Mesh-1, NN-2	10	3	1	1	34	17	35
	NN-1, Mesh-2	9	3	1	1	29	15	32

1. GPR file with 4-in/4-out ports.
2. Clustered GPR file similar to the one described in [202] with 4- in/4-out ports.
3. GPR file with 4-in/4-out ports, and an additional 16 Internal Registers (IRs) accessible from FPGA. ISE1 through ISE4 use these registers to communicate the intermediate values.
4. GPR file and a block of 8×8 IRs accessible from FPGA. At the beginning of IDCT calculation, the entire 8×8 data-block is moved from memory to this register file. At the end of calculation, the block is moved back to memory.

Since the second ISE set contains a single ISE with 8-inputs and 8-outputs, a GPR file with 8-in/8-out ports is chosen. The 8×8 IR file was also put inside the FPGA for the second set. In total, 45 different design points for the first and second ISE sets are explored.

As can be easily seen from Table 8.18, the IDCT-2 cluster topology and {NN-1, Mesh-2} inter-cluster connectivity results in the lowest critical path for all ISEs. The IDCT-3 cluster topology, in spite of having more PEs than IDCT-2, do not give better critical paths or higher cluster utilization. The IDCT-1 topology is smaller in area than IDCT-2, but results in longer critical paths. Table 8.18 also shows that the first set of ISEs (ISE1 through ISE4) always result in smaller number of execution cycles than the second set. Moreover, the second set requires more data bandwidth from the GPR file. The speed-up, register file area and cluster area results for the case study are summarized in Fig. 8.13. For the sake of simplicity, only the results

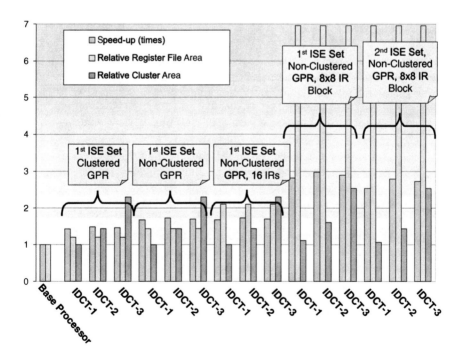

Fig. 8.13 Results for IDCT case study

for {NN-1, Mesh-2} inter-cluster connectivity scheme is presented, since it always produced the best result. As the figure clearly shows, the design space is highly non-linear. As far as speed-up is concerned, the combination of IDCT-2 FPGA structure, 4-in/4-out register file with 8×8 IR block, and the first set of ISEs produce the best result. However, the 8×8 IR block significantly increases the register file area. Similarly, the IDCT-1 cluster topology produces almost as good speed-up results as the IDCT-2, but it is significantly smaller in area. As a consequence, the inclusion of the IDCT-2 topology and 64 IRs in the final design might depend on the area constraints. Such non-linearities of the design space underlines the importance of architecture exploration.

8.6.2 JPEG IDCT and DCT Kernels

The JPEG IDCT and DCT kernels are very similar, but not exactly the same, to the MPEG2 IDCT kernel. The JPEG IDCT kernel uses an extra quantization table and requires even more data-bandwidth from the base processor. Additionally, the JPEG DCT/IDCT kernels require more FPGA clusters due to the presence of larger number of operators. As is the general case with all IDCT/DCT kernels, the 8×8 IR block in the re-configurable fabric results in better performance for JPEG IDCT/DCT.

8.6.3 SAD Kernel

The SAD kernel calculates the difference of two vectors, element by element, and sums up the absolute value of these differences. This kernel is extensively used in motion estimation blocks of different video compression algorithm like MPEG2, H.263 and H.264. Since various implementations of SAD differ very little, a set of ISEs which can be reused over all of them is conceived. The FPGA topologies explored for SAD are marked as SAD-1, SAD-2 and SAD-3 in Fig. 8.14. Each of these FPGA topologies included a special PE, named AB_ALU, which can calculate $abs(x - y)$ or $abs(x + y)$ apart from common arithmetic and logic operations. The $abs(x - y)$ was included because the SAD kernel repeatedly uses this function. The utility of $abs(x + y)$ is explained later. It is also found that the high degree of parallelism available in the SAD kernel can not be exploited through the GPR file. When the ISEs requiring 8 inputs/4 outputs into SAD code are inserted, they caused large amount of GPR spills. However, the performance improved greatly when 16 IRs were provided to SAD ISEs for communication.

8.6.4 SHTD Kernel

The SHTD kernel from H.264 uses several additions and subtractions, $abs(x - y)$ operations, and $abs(x + y)$ operations. Moreover, the software code for the kernel

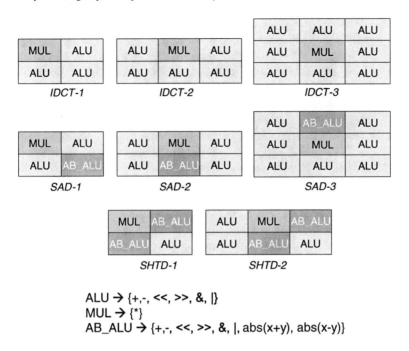

ALU → {+,-, <<, >>, &, |}
MUL → {*}
AB_ALU → {+,-, <<, >>, &, |, abs(x+y), abs(x-y)}

Fig. 8.14 Cluster structures for different applications

uses two 4×4 local arrays. Considerable speed-up can be achieved if 32 IRs are used instead of these local arrays. The SHTD-1 and SHTD-2 FPGA topologies in Fig. 8.14 were used in the FPGA exploration of the SHTD kernel.

The best results (considering the achievable speed-up only) for all the kernels are presented in Table 8.19. The FPGA configurations explored for each one of them are shown in Fig. 8.14. For the first three applications namely, MPEG2 IDCT, JPEG DCT and JPEG IDCT, IDCT-2 cluster configuration is used. For SAD and SHTD, SAD-2 and SHTD-2 cluster configurations are used respectively. The connectivity style NN-1, Mesh-2 yielded the best results for all the

Table 8.19 Best configuration of application kernels

Application	Storage interface				FPGA structure		Base processor area (KGates)	Speed-up (times)
	GPR I/O	IR Count	IR I/O	Area (KGates)	Cluster count	Cluster area (KGates)		
MPEG2 IDCT	4/4	64	8/4	76.16	27	25.63	96.35	3
JPEG DCT	4/4	64	8/4	76.16	32	25.63	96.35	4.45
JPEG IDCT	4/4	64	8/4	76.16	30	25.63	96.35	2.08
SAD	8/4	16	4/4	32.14	29	26.52	51.77	2.26
SHTD	8/4	32	4/4	48.93	16	29.72	68.56	5.59

applications shown above. One can see that different design points yield best results for different applications. The best cluster topologies of all the different kernels can be unified in the SHTD-2 structure. Similarly, the best interface setting is to have 64 IRs with 8-inputs/4-outputs and a GPR file with 4-inputs/4-outputs. The 64 IRs can be used in IDCT/DCT for storing the 8×8 data block, or for inter ISE communication in SAD, or for storing the local arrays in SHTD kernel. This case study points out how important architecture exploration is for the rASIP design process. Using the proposed design flow, 45 different design points for MPEG2 IDCT and more than 60 different design points for the other kernels taken together, are explored. This exploration was done within 5 man days which would be extremely difficult without the comprehensive tool support offered by the proposed methodology. The results also clearly advocate a careful investigation of various rASIP design alternatives for a variety of applications. For example, a FPGA designed for IDCT/DCT is not good enough for SAD/SHTD, since it does not contain the *abs* operator. Similarly, the number of internal registers required for SHTD (32 registers) is not enough for the DCT/IDCT kernel.

Chapter 9
Past, Present and Future

We are made wise not by the recollection of our past, but by the responsibility for our future.
George Bernard Shaw, Author, 1856–1950

9.1 Past

Since the inception of modern era of communication and information technology, the whole world is witnessing rapid advancement in the field of digital system design. Starting from a few transistors on a single integrated circuit conceived during late 1950s, we are now in an era of billions of transistors integrated to form heterogeneous System-on-Chip (SoC). Albeit, the drive to squeeze more and more elements in a single system is coming with diminishing returns mostly due to the challenges posed by energy efficiency and skyrocketing Non-Recurring Engineering (NRE) costs. Therefore, a wide range of system components are now being offered as the building blocks of modern SoC to balance the conflicting demands of flexibility (to avoid large NRE cost) and performance.

The range of the system components are huge, given the unique offering of each of them in terms of area efficiency, energy efficiency and flexibility. The most flexible system components are usually the least energy efficient ones e.g. general purpose processors, whereas with highest energy efficiency comes at the cost of minimum flexibility. Naturally, the focus of system designers are on those components, which can be engineered to strike a balance between the conflicting requirements of performance and flexibility. ASIPs as well as FPGAs featured in this group and thus received strong attention of designers over the past decade [17, 28, 29, 203].

Inevitably, designers toyed with the idea of merging two design styles namely, fixed and re-configurable logic, together into one single application-specific processor. A wide range of processors, thus formed, are proposed over the years. Clearly, the number of choices for different design decisions are staggering (refer Fig. 9.1). Therefore novel design concepts, which limit these design choices, for partially re-configurable processors kept on appearing in the research community [35, 66, 33]. Even after years of research focus, partially re-configurable systems are yet to catch the fancy of commercial vendors. Few issues, until recently, remained challenging enough to hinder a greater acceptability of these systems. These are as following.

- It was never easy to design a processor at first place. A processor design is basically to decide the partitioning of hardware and software, to design these two interrelated parts and then to design tools which enable easy mapping from the

A. Chattopadhyay et al., *Language-driven Exploration and Implementation of Partially Re-configurable ASIPs,* DOI 10.1007/978-1-4020-9297-8_9,
© Springer Science+Business Media B.V. 2009

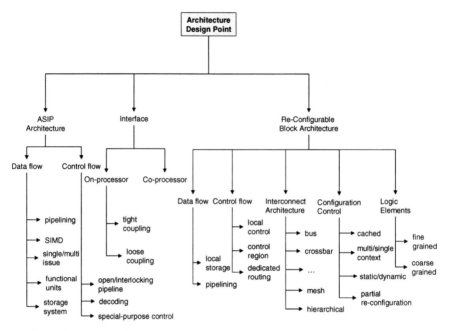

Fig. 9.1 rASIP design choices

software to the hardware. The overall process involves detailed understanding of
the influence of each design point on the rest of the design. With the invention of
ADLs [17, 16, 11], the processor design complexity is much more manageable
now.

- FPGAs are traditionally viewed as a prototyping platform. Its huge advantage
  as a re-configurable component is put off by the disadvantages of designing an
  FPGA from scratch (specifically due to its highly optimized physical design) and
  its poor performance compared to dedicated ASIC or even processor in case of
  control-dominated circuits. Of late, designers have suggested a more application-
  specific model of re-configurable architectures. These architectures feature dedicated functional blocks (instead of generic LUT) and less flexibility in routing
  architecture [41, 184] compared to the that found in fine-grained FPGAs.
- The aforementioned two difficulties, when faced together, makes the task even
  more challenging. Most of the partially re-configurable processor, therefore,
  shied away from attempting a generic design methodology and restricted the
  problem within a specific template proposed of their own. Obviously, that pre-
  vented application-specific design space exploration, which is much needed in
  view of today's fast changing applications.

With the emergence of new design tools and new design points in both re-configurable
architectures and processors, the present times seemed ripe for a strong re-entry of
partially re-configurable processor design. This time, equipped with powerful and
generic design tools.

9.2 Present

In view of emerging language-driven processor design concept, few attempts at modelling partially re-configurable processors at high-level of abstraction are made [34, 46]. Though these models allow early design space exploration, a generic processor design paradigm is lacking in both of these. While the ADRES architecture [34] advocates a VLIW-like structure, the Pleiades [46] has a pre-defined system template – both allowing limited designer freedom.

The work presented in this book started with the identification of a broad range of possible design choices in the field of partially re-configurable processor design. Thenceforth, the design flow is segmented into two subsequent phases namely, pre-fabrication phase and post-fabrication phase. At this point, the exact requirements of design tools are noted. To facilitate fast design space exploration and convenient modelling – a generic language-driven platform for ASIP design [17] is chosen as the basis. The language is extended with new keywords and sections to enable the modelling of the entire partially re-configurable processor. This necessitated a re-working of the existing tool flow as well as developing new sets of tools. Furthermore, the state-of-the-art research tools from processor design automation are integrated in the proposed partially re-configurable processor design framework. In the following, the basic design principles and associated tools are summarized.

9.2.1 Pre-Fabrication Design Flow

The Fig. 9.2 captures the pre-fabrication rASIP design flow graphically. The design starts from the application profiling, which is done via static and dynamic profiling [113] of the application in an architecture-independent manner. The profiling helps to narrow down the architectural design space. With the aid of the profiler, the decisions concerning the memory hierarchy, number of registers, processor architecture (e.g. RISC, VLIW) can be taken. These decisions are formally described using the language LISA 3.0, which is used as the modelling platform for complete rASIP. An associated tool, termed *Coding Leakage Explorer*, analyzes the coding contribution of different LISA operations and determines the free coding space a.k.a *coding leakage*. The higher the coding leakage in a particular LISA operation, the more number of special instructions it can accommodate as its children. The rASIP software tools e.g. simulator, assembler, linker, profiler are automatically derived from the LISA description. These set of software tools allow the designer to do a target-specific application profiling. With the aid of the LISA simulator and coding leakage explorer, the partitioning of data flow as well as control flow can be performed by the designer. The partitioning needs to consider the future evolution of target application(s). Accordingly, room for future instructions, storage space, interface bandwidth etc. must be reserved. The LISA-based RTL synthesis tool automatically generates the RTL description of the base processor and the re-configurable block as per the designer-directed partitioning. The partitioning generates the interface between the processor

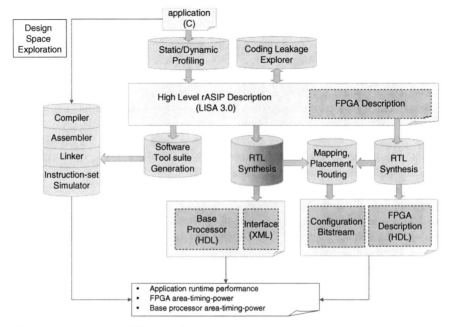

Fig. 9.2 Pre-fabrication rASIP design flow

and the re-configurable block automatically. The complete interface is stored in the XML format. The storage of the interface is necessary in order to ensure that the interface restrictions are not violated during the post-fabrication rASIP enhancements. The generated HDL description for the base processor can be further synthesized using commercial gate-level synthesis flows [139]. For the re-configurable part to map, the detailed structural specification of the coarse-grained FPGA can be formally described in LISA. The re-configurable part of the processor is internally taken from the RTL synthesis tool and then mapped, placed and routed on to the coarse-grained FPGA as per the LISA specification. The output of this is a configuration bitstream, which is passed on to the RTL description of the coarse-grained FPGA for simulation.

The design decisions to be taken in the pre-fabrication phase are as following.

- rASIP data flow and partitioning.
- rASIP control flow and partitioning.
- rASIP ISA.
- Base processor micro-architecture.
- Re-configurable block micro-architecture.

These decisions are key to play the trade-off between the flexibility and the performance of the rASIP. For example, the availability of local decoding inside the re-configurable block allows flexibility in adding custom instructions. On the other hand, the decoder structure, being irregular in nature, adds to the interconnect cost

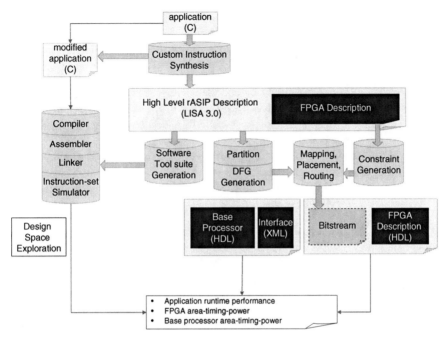

Fig. 9.3 Post-fabrication rASIP design flow

and power budget of the re-configurable block. The decisions taken during the pre-fabrication design flow constrain the optimization benefits (or in other words, the design space) achievable during the post-fabrication design. For example, the interface between the fixed processor part and the re-configurable block is decided during the pre-fabrication flow. This decision cannot be altered once the processor is fabricated.

9.2.2 Post-fabrication Design Flow

The post-fabrication rASIP design flow is shown in Fig. 9.3. In this phase, the rASIP is already fabricated. Due to this, several parts of the design cannot be altered (labelled dark). The major design decision, which needs to be taken, is the selection and synthesis of custom instructions to reduce the application runtime. Consequently, the custom instructions need to be mapped, placed and routed on the coarse-grained FPGA.

In the proposed flow, the custom instruction synthesis tool (ISEGen), presented in [113], is used. This tool is able to accept generic interface constraints and produce custom instructions in LISA description format. Apart from that, manual custom instruction / datapath modelling is also possible via LISA description in the post-fabrication phase. The only restriction is that the pre-fabrication design constraints must be adhered to by the designer. This naturally indicates that several parts (base processor, coarse-grained FPGA structure) of the LISA description cannot be altered. The post-fabrication design constraints can be summarized as following.

- The special-purpose or custom instructions must not violate the existing interface between re-configurable block and the base processor. Actually, the custom instruction synthesis tool used in the presented flow accepts this interface as an input constraint. Even after that, during the RTL synthesis phase, the fabricated interface is matched with the new generated interface.
- The additional number of instructions must not exceed the number allowed by the free coding space available.
- The overall area of the custom data path should be within the area budget of the existing re-configurable block.

In the post-fabrication design flow, ISEGen generates the custom instructions with potentially high speed-up. The calls to these custom instructions are embedded in the application. The behavior of the custom instructions are generated in form of LISA operations, which can be directly plugged in to the existing LISA model. The software tool suite for the modified rASIP description is generated automatically. The re-configurable block, as in pre-fabrication flow, is subjected to the FPGA mapping, placement and routing flow, with a focus on area constraints. Essentially, the same set of tools are used in both pre- and post-fabrication rASIP design.

9.3 Future

The design flow proposed in this book made an important step towards generic rASIP design space exploration and implementation. Continued interest on this topic depends heavily on the emergence of more demanding applications as well as aggressive requirement of energy-efficiency – both of which seems highly probable at present. Various major research endeavors can, thus, be undertaken around the current rASIP design framework. These are presented categorically here.

Application Analysis Definitely, it is the application developer, who decides best among the design choices. However, it is difficult for the application developer to ascertain the design choice when the system is large and complex – as in rASIP. On the other hand, from the perspective of system developers – it is imperative to have useful tools to understand the static and dynamic behavior of the application. The current set of tools for high-level application analysis includes target-independent profiling [105] and target-based simulation. Target-independent profiling tools still lack the cognizance of an experienced hardware designer. For example, a poor software implementation of barrel shifter can hardly be tackled automatically by a profiling/synthesis tool. This is more true in case of FPGA-based architectures as the target. The user of profiling tool expects to have early design suggestions on how to group the basic blocks, what should be the granularity or what kind of routing architecture is most suitable. These questions become more and more important with the idea of having a application-specific FPGA in the system.

Accommodating Further FPGA Design Points The approach suggested in this book for generic modelling coarse-grained FPGAs is a step towards the future of re-configurable architecture. Current world of re-configurable architectures lack generic high-level modelling platform, generic synthesis tools and physically optimized implementation route. While the first two problems are addressed in this book, the last one is yet undone. Researchers in future will definitely have to find a mechanism to define an implementation methodology (library-based, perhaps) to open up the FPGA design to a greater community.

In the area of high-level FPGA modelling, the description style proposed in this book can be a starting point. It can be enhanced to permit dedicated channels, switching network and such common fine-grained FPGA features. Though there is a scope of lively argument over the necessity of detailed routing structures in coarse-grained application-specific FPGA, yet a generic language should be able to accommodate more than what is covered in this book. In tune with these extensions, the synthesis tools also need to be updated without losing the genericity.

Accurate Cost Estimation Fast downscaling of technology has extended the horizon of EDA researchers tremendously. Error-tolerant design, temperature-aware synthesis are increasingly common words in low-level synthesis. With the cost of fabrication increasing, it will be impossible for future designers to ignore these effects even during high-level design. To equip the proposed design flow considering such prospects, it is important to accurately estimate the energy-area-timing of a rASIP in high-level itself. It is also interesting to estimate the advantage/disadvantage of rASIP against FPGA and ASIP across scaling technology.

System-level Integration With the currently proposed rASIP design framework, it is possible to view the rASIP as another system component and build a SoC with it. An interesting work in this direction can be to represent a complete SoC – with ASIC, ASIP, micro-processor, re-configurable hardware blocks, communication network – in high-level abstraction of LISA. Except the communication network, and for some highly optimized ASIC, the rest can be modelled very well in LISA. It is conjectured that, such a representation will increase the design productivity significantly.

References

1. Moore's Law. *http://www.intel.com/technology/mooreslaw/*.
2. Joerg Henkel. Closing the SoC Design Gap. *Computer*, 36(9):119–121, 2003.
3. G. M. Amdahl, G. A. Blaauw and F. P. Brooks, Jr. Architecture of the IBM System/360. *IBM Journal of Research and Development*, 8(2), 1964.
4. J. Markoff. Shift to simplicity promises big advances in computing. (reduced instruction set computers). *New York Times*, 137(47), 1988.
5. J. Cocke and V. Markstein. The Evolution of RISC Technology at IBM. *IBM Journal of Research and Development*, 34(1), 1990.
6. David A. Patterson and Carlo H. Sequin. A VLSI RISC. *Computer*, 15(9):8–21, 1982.
7. T. Makimoto. The Rising Wave of Field Programmability. In *FPL '00: Proceedings of The Roadmap to Reconfigurable Computing, 10th International Workshop on Field-Programmable Logic and Applications*, pages 1–6, London, UK, 2000. Springer-Verlag.
8. T. Gloekler, S. Bitterlich, and H. Meyr. ICORE: A Low-Power Application Specific Instruction Set Processor for DVB-T Acquisition and Tracking. In *Proc. of the ASIC/SOC conference*, 2000.
9. Texas Instruments. *Digital Media Processor. http://www.ti.com/*.
10. A. Hoffmann, H. Meyr and R. Leupers. *Architecture Exploration for Embedded Processors with LISA*. Kluwer Academic Publishers, 2002.
11. P. Grun, A. Halambi, A. Khare, V. Ganesh, N. Dutt, and A. Nicolau. EXPRESSION: An ADL for System Level Design Exploration. Technical report, Department of Information and Computer Science, University of California, Irvine, 1998.
12. A. Fauth, J. Van Praet, and M. Freericks. Describing Instruction Set Processors Using nML. In *Proc. of the European Design and Test Conference (ED&TC)*, 1995.
13. G. Hadjiyiannis, S. Hanono, and S. Devadas. ISDL: An Instruction Set Description Language for Retargetability. In *Proc. of the Design Automation Conference (DAC)*, 1997.
14. M. Itoh and S. Higaki and J. Sato and A. Shiomi and Y. Takeuchi and A. Kitajima and M. Imai. PEAS-III: An ASIP Design Environment. In *Proc. of the Int. Conf. on Computer Design (ICCD)*, 2000.
15. P. Mishra and N. Dutt. *Processor Description Languages*. Morgan Kaufmann Publishers, 2008.
16. Target Compiler Technologies. *http://www.retarget.com*.
17. CoWare/Processor Designer. *http://www.coware.com*.
18. UC Irvine/EXPRESSION. *http://www.ics.uci.edu/~express/index.htm*.
19. JJ. Ceng, M. Hohenauer, R. Leupers, G. Ascheid, H. Meyr and G. Braun. C Compiler Retargeting Based on Instruction Semantics Models. In *DATE*, Munich, Germany, March 2005.
20. M. Hohenauer, H. Scharwaechter, K. Karuri, O. Wahlen, T. Kogel, R. Leupers, G. Ascheid, H. Meyr, G. Braun and H. van Someren. A Methodology and Tool Suite for C Compiler Generation from ADL Processor Models. In *Proceedings of the conference on Design, Automation & Test in Europe (DATE)*, Paris, France, Feb 2004.

21. P. Mishra and N. Dutt. Automatic functional test program generation for pipelined processors using model checking. In *Seventh IEEE International Workshop on High Level Design Validation and Test (HLDVT)*, 2002.

22. P. Mishra, N. Dutt, N. Krishnamurthy and M. S. Abadir. A Top-Down Methodology for Validation of Microprocessors. In *IEEE Design and Test of Computers (Design and Test)*, 2004.

23. O. Luethje. A Methodology for Automated Test Generation for LISA Processor Models. In *The Twelfth Workshop on Synthesis and System Integration of Mixed Information Technologies, Kanazawa, Japan*, October 18–19, 2004.

24. A. Chattopadhyay, A. Sinha, D. Zhang, R. Leupers, G. Ascheid and H. Meyr. Integrated Verification Approach during ADL-driven Processor Design. In *IEEE International Workshop on Rapid System Prototyping (RSP)*, Chania, Crete, June 2006.

25. P. Mishra and N. Dutt. Functional coverage driven test generation for validation of pipelined processors. In *Design Automation and Test in Europe (DATE)*, pages 678–683, 2005.

26. A. Chattopadhyay, A. Sinha, D. Zhang, R. Leupers, G. Ascheid, H. Meyr. ADL-Driven Test Pattern Generation for Functional Verification of Embedded Processors. In *12th IEEE European Test Symposium*, May 21–24, 2007.

27. R. Gonzales. Xtensa: A configurable and extensible processor. *IEEE Micro*, 2000.

28. ARC. *http://www.arc.com*.

29. Tensilica. *http://www.tensilica.com*.

30. M. Puig-Medina, G. Ezer and P. Konas. Verification of Configurable Processor Cores. In *DAC '00: Proceedings of the 37th conference on Design automation*, pages 426–431, New York, NY, USA, 2000. ACM Press.

31. R. E. Bryant. Graph-based Algorithms for Boolean Function Manipulation. *IEEE Transactions on Computers*, 35(8):677–691, 1986.

32. G. D. Micheli. *Synthesis and Optimization of Digital Circuits*. McGraw-Hill, 1 edition. ISBN: 0070163332.

33. Stretch. *http://www.stretchinc.com*.

34. B. Mei, A. Lambrechts, D. Verkest, J. Mignolet and R. Lauwereins. Architecture Exploration for a Reconfigurable Architecture Template. *IEEE Design and Test*, 22(2):90–101, 2005.

35. P. M. Athanas and H. F. Silverman. Processor Reconfiguration Through Instruction-Set Metamorphosis. *IEEE Computer*, 26(3):11–18, 1993.

36. D. A. Buell, J. M. Arnold, and W. J. Kleinfelder. *Splash 2: FPGAs in a Custom Computing Machine*. IEEE Computer Society Press, 1996.

37. W. S. Carter, K. Duong, R. H. Freeman, H. Hsieh, J. Y. Ja, J. E. Mahoney, L. T. Ngo and S. L. Sze. A User Programmable Reconfigurable Logic Array. *In Proceedings of IEEE Custom Integrated Circuits Conference*, pages 233–235, 1986.

38. S. Brown and J. Rose. FPGA and CPLD Architectures: A Tutorial. *IEEE Design and Test*, 13(2):42–57, 1996.

39. Xilinx. *http://www.xilinx.com*.

40. Altera. *http://www.altera.com/*.

41. MathStar. *http://www.mathstar.com/*.

42. A. DeHon. The Density Advantage of Configurable Computing. *Computer*, 33(4):41–49, 2000.

43. D. Pham, H. Anderson, E. Behnen, M. Bolliger, S. Gupta, P. Hofstee, P. Harvey, C. Johns, J. Kahle, A. Kameyama, J. Keaty, B. Le, S. Lee, T. Nguyen, J. Petrovick, M. Pham, J. Pille, S. Posluszny, M. Riley, J. Verock, J. Warnock, S. Weitzel and D. Wendel. Key Features of the Design Methodology enabling a multi-core SoC Implementation of a first-generation CELL Processor. In *ASP-DAC '06: Proceedings of the 2006 conference on Asia South Pacific Design automation*, pages 871–878, New York, NY, USA, 2006. ACM Press.

44. International Technology Roadmap for Semiconductors. *System Drivers - 2006 Update http://www.itrs.net/*.

45. International Technology Roadmap for Semiconductors. *Design - 2006 Update http://www.itrs.net/*.

46. M. Wan, H. Zhang, V. George, M. Benes, A. Abnous, V. Prabhu and J. Rabaey. Design Methodology of a Low-Energy Reconfigurable Single-Chip DSP System. *Journal of VLSI Signal Processing Systems*, 28(1–2):47–61, 2001.

47. S. C. Goldstein, H. Schmit, M. Moe, M. Budiu, S. Cadambi, R. Reed Taylor and R. Laufer. PipeRench: A Coprocessor for Streaming multimedia Acceleration. In *ISCA*, pages 28–39, 1999.

48. F. Barat, R. Lauwereins and G. Deconinck. Reconfigurable Instruction Set Processors from a Hardware/Software Perspective. *IEEE Transactions on Software Engineering*, 28(9):847–862, 2002.

49. K. Compton and S. Hauck. Reconfigurable Computing: a Survey of Systems and Software. *ACM Computing Survey*, 34(2):171–210, 2002.

50. J. R. Hauser and J. Wawrzynek. Garp: a MIPS Processor with a Reconfigurable Coprocessor. In *Proceedings of the 5th IEEE Symposium on FPGA-Based Custom Computing Machines (FCCM '97)*, page 12. IEEE Computer Society, 1997.

51. E. M. Panainte, S. Vassiliadis, S. Wong, G. Gaydadjiev, K. Bertels and G. Kuzmanov. The MOLEN Polymorphic Processor. *IEEE Transactions on Computers*, 53(11):1363–1375, 2004.

52. M. Wirthlin and B. Hutchings. A Dynamic Instruction Set Computer. In Peter Athanas and Kenneth L. Pocek, editors, *IEEE Symposium on FPGAs for Custom Computing Machines*, pages 99–107, Los Alamitos, CA, 1995. IEEE Computer Society Press.

53. J. M. Arnold, D. A. Buell and E. G. Davis. Splash 2. In *SPAA '92: Proceedings of the fourth annual ACM symposium on Parallel Algorithms and Architectures*, pages 316–322, New York, NY, USA, 1992. ACM Press.

54. Sun Microsystems. *http://www.sun.com/processors/*.

55. M. Wazlowski, L. Agarwal, T. Lee, A. Smith, E. Lam, P. Athanas, H. Silverman and S. Ghosh. PRISM-II compiler and architecture. In Duncan A. Buell and Kenneth L. Pocek, editors, *IEEE Workshop on FPGAs for Custom Computing Machines*, pages 9–16, Los Alamitos, CA, 1993. IEEE Computer Society Press.

56. D.C. Chen and J.M. Rabaey. A Reconfigurable Multiprocessor IC for Rapid Prototyping of Algorithmic-specific High-speed DSP Data Paths. *IEEE Journal of Solid-State Circuits*, 27(12):1895–1904, 1992.

57. A.K. Yeung and J.M. Rabaey. A 2.4 GOPS data-driven reconfigurable multiprocessor IC for DSP. *42nd ISSCC, IEEE International Solid-State Circuits Conference, 1995. Digest of Technical Papers*, pages 108–109, 1995.

58. C. Iseli and E. Sanchez. Spyder: a SURE (SUperscalar and REconfigurable) Processor. *Journal of Supercomputing*, 9(3):231–252, 1995.

59. M. J. Wirthlin, B. L. Hutchings and K. L. Gilson. The Nano Processor: A Low Resource Reconfigurable Processor. In Duncan A. Buell and Kenneth L. Pocek, editors, *IEEE Workshop on FPGAs for Custom Computing Machines*, pages 23–30, Los Alamitos, CA, 1994. IEEE Computer Society Press.

60. R. Razdan and M. D. Smith. A High-Performance Microarchitecture with Hardware-Programmable Functional Units. In *Proceedings of the 27th Annual International Symposium on Microarchitecture*, pages 172–80, 1994.

61. R. Wittig and P. Chow. OneChip: An FPGA processor with reconfigurable logic. In Kenneth L. Pocek and Jeffrey Arnold, editors, *IEEE Symposium on FPGAs for Custom Computing Machines*, pages 126–135, Los Alamitos, CA, 1996. IEEE Computer Society Press.

62. J. A. Jacob and P. Chow. Memory Interfacing and Instruction Specification for Reconfigurable Processors. In *Proceedings of the 1999 ACM/SIGDA Seventh International Symposium on Field Programmable gate arrays*, pages 145–154. ACM Press, 1999.

63. C. Ebeling, D. C. Cronquist and P. Franklin. RaPiD - Reconfigurable Pipelined Datapath. In *Proceedings of the 6th International Workshop on Field-Programmable Logic, Smart Applications, New Paradigms and Compilers*, pages 126–135. Springer-Verlag, 1996.

64. C. Ebeling, C. Fisher, G. Xing, M. Shen and H. Liu. Implementing an OFDM Receiver on the RaPiD Reconfigurable Architecture. *IEEE Transactions on Computers*, 53(11):1436–1448, 2004.

65. E. Mirsky and A. DeHon. MATRIX: A Reconfigurable Computing Architecture with Configurable Instruction Distribution and Deployable Resources. In Kenneth L. Pocek and Jeffrey Arnold, editors, *IEEE Symposium on FPGAs for Custom Computing Machines*, pages 157–166, Los Alamitos, CA, 1996. IEEE Computer Society Press.

66. E. Waingold, M. Taylor, D. Srikrishna, V. Sarkar, W. Lee, V. Lee, J. Kim, M. Frank, P. Finch, R. Barua, J. Babb, S. Amarasinghe and A. Agarwal. Baring it all to Software: Raw Machines. *Computer*, 30(9):86–93, 1997.

67. T. J. Callahan, J. R. Hauser and J. Wawrzynek. The Garp Architecture and C Compiler. *Computer*, 33(4):62–69, 2000.

68. T. J. Callahan, P. Chong, A. DeHon and J. Wawrzynek. Fast Module Mapping and Placement for Datapaths in FPGAs. In *FPGA '98: Proceedings of the 1998 ACM/SIGDA sixth international symposium on field programmable gate arrays*, pages 123–132, New York, NY, USA, 1998. ACM Press.

69. C. R. Rupp, M. Landguth, T. Garverick, E. Gomersall, H. Holt, J. M. Arnold and M. Gokhale. The NAPA Adaptive Processing Architecture. In *Proceedings of the IEEE Symposium on FPGAs for Custom Computing Machines*, page 28. IEEE Computer Society, 1998.

70. J. M. Arnold. An Architecture Simulator for National Semiconductor's Adaptive Processing Architecture (NAPA). In *IEEE Symposium on FPGAs for Custom Computing Machines*, pages 271–272, 1998.

71. T. Miyamori and K. Olukotun. REMARC (abstract): Reconfigurable Multimedia Array Coprocessor. In *Proceedings of the 1998 ACM/SIGDA sixth international symposium On Field programmable gate arrays*, page 261. ACM Press, 1998.

72. MIPS. *MIPS Technologies. http://www.mips.com/*.

73. P. Graham and B. E. Nelson. Reconfigurable Processors for High-Performance, Embedded Digital Signal Processing. In *Proceedings of the 9th International Workshop on Field-Programmable Logic and Applications*, pages 1–10. Springer-Verlag, 1999.

74. S. Hauck, T. W. Fry, M. M. Hosler and J. P. Kao. The Chimaera Reconfigurable Functional Unit. *IEEE Transactions on Very Large Scale Integrated Systems*, 12(2):206–217, 2004.

75. Z. Alex Ye, A. Moshovos, S. Hauck and P. Banerjee. CHIMAERA: a High-performance Architecture with a tightly-coupled Reconfigurable Functional Unit. In *ISCA*, pages 225–235, 2000.

76. GNU. *gcc, gprof, gcov. http://www.gnu.org/software/*.

77. T. Austin, E. Larson and D. Ernst. SimpleScalar: An Infrastructure for Computer System Modeling. *Computer*, 35(2):59–67, 2002.

78. H. Singh, M. Lee, G. Lu, F. J. Kurdahi, N. Bagherzadeh and E. M. Chaves Filho. MorphoSys: An Integrated Reconfigurable System for Data-Parallel and Computation-Intensive Applications. *IEEE Transactions on Computers*, 49(5):465–481, 2000.

79. M. W. Hall, J. M. Anderson, S. P. Amarasinghe, B. R. Murphy, S. Liao, E. Bugnion and M. S. Lam. Maximizing Multiprocessor Performance with the SUIF Compiler. *Computer*, 29(12):84–89, 1996.

80. S. Majzoub, R. Saleh, and H. Diab. Reconfigurable Platform Evaluation Through Application Mapping and Performance Analysis. In *IEEE International Symposium on Signal Processing and Information Technology*, pages 496–501, 2006.

81. G. J. M. Smit, P. J. M. Havinga, L. T. Smit, P. M. Heysters and M. A. J. Rosien. Dynamic Reconfiguration in Mobile Systems. In *FPL '02: Proceedings of the Reconfigurable Computing Is Going Mainstream, 12th International Conference on Field-Programmable Logic and Applications*, pages 171–181, London, UK, 2002. Springer-Verlag.

82. M. A. J. Rosien, Y. Guo, G. J. M. Smit and T. Krol. Mapping Applications to an FPFA Tile. In *DATE*, pages 11124–11125, 2003.

83. R. Enzler and M. Platzner. Application-driven design of dynamically reconfigurable processors. Technical Report, Swiss Federal Institute of Technology (ETH) Zurich, Electronics Laboratory, 2001.

84. C. Plessl and M. Platzner. Zippy - A Coarse-grained Reconfigurable Array with Support for Hardware Virtualization. In *ASAP '05: Proceedings of the 2005 IEEE International Conference on Application-Specific Systems, Architecture Processors (ASAP'05)*, pages 213–218, Washington, DC, USA, 2005. IEEE Computer Society.

85. A. Lodi, M. Toma, F. Campi, A. Cappelli, R. Canegallo and R. Guerrieri. A VLIW Processor with Reconfigurable Instruction Set for Embedded Applications. *IEEE Journal of Solid-State Circuits*, 38(11):1876–1886, 2003.

86. A. Nohl and G. Braun and O. Schliebusch and R. Leupers, H. Meyr and A. Hoffmann. A Universal Technique for Fast and Flexible Instruction-set Architecture Simulation. In *DAC '02: Proceedings of the 39th conference on Design automation*. ACM Press, 2002.

87. C. Mucci, F. Campi, A. Deledda, A. Fazzi, M. Ferri, M. Bocchi. A Cycle-Accurate ISS for a Dynamically Reconfigurable Processor Architecture. In *Proceedings of the 19th IEEE International Parallel and Distributed Processing Symposium (IPDPS'05)*, 2005.

88. IBM. *PowerPC. http://www-03.ibm.com/chips/power/powerpc/*.

89. E. Moscu Panainte, K. Bertels and S. Vassiliadis. The Molen Compiler for Reconfigurable Processors. *ACM Transactions on Embedded Computing Systems*, 6(1), 2007.

90. E. Moscu Panainte, K. Bertels and S. Vassiliadis. Compiler-driven FPGA-area Allocation for Reconfigurable Computing. In *Proceedings of Design, Automation and Test in Europe 2006 (DATE 06)*, pages 369–374, March 2006.

91. T. von Sydow, B. Neumann, H. Blume and T. G. Noll. Quantitative Analysis of Embedded FPGA-Architectures for Arithmetic. In *Proceedings of the 17th International Conference on Application-specific Systems, Architectures and Processors (ASAP'06)*, volume 0, pages 125–131, Los Alamitos, CA, USA, 2006. IEEE Computer Society.

92. T. von Sydow, M. Korb, B. Neumann, H. Blume and T. G. Noll. Modelling and Quantitative Analysis of Coupling Mechanisms of Programmable Processor Cores and Arithmetic Oriented eFPGA Macros. In *Proceedings of the IEEE International Conference on Reconfigurable Computing and FPGA's, ReConFig 2006*, pages 1–10, 2006.

93. B. Mei, S. Vernalde, D. Verkest and R. Lauwereins. Design Methodology for a Tightly Coupled VLIW/Reconfigurable Matrix Architecture: A Case Study. In *DATE 2004: Proceedings of the conference on Design, automation and test in Europe*, 2004.

94. P. Op de Beeck, F. Barat, M. Jayapala and R. Lauwereins. CRISP: A Template for Reconfigurable Instruction Set Processors. In *FPL '01: Proceedings of the 11th International Conference on Field-Programmable Logic and Applications*, pages 296–305, London, UK, 2001. Springer-Verlag.

95. The Impact Research Group. *http://www.crhc.uiuc.edu/Impact/*.

96. B. Mei, S. Vernalde, D. Verkest, H. Man and R. Lauwereins. DRESC: A Retargetable Compiler for Coarse-grained Reconfigurable Architectures. In *International Conference on Field Programmable Technology*, 2002.

97. M. S. Lam. Software Pipelining: An Effective Scheduling Technique for VLIW Machines. *SIGPLAN Notices*, 39(4):244–256, 2004.

98. R. Hartenstein, M. Herz, T. Hoffmann and U. Nageldinger. KressArray Xplorer: a new CAD environment to optimize Reconfigurable Datapath Array. In *Proceedings of the 2000 conference on Asia South Pacific Design Automation*.

99. R. W. Hartenstein and R. Kress. A Datapath Synthesis System for the Reconfigurable Datapath Architecture. In *ASP-DAC '95: Proceedings of the 1995 conference on Asia Pacific Design Automation*, page 77, New York, NY, USA, 1995. ACM Press.

100. R. Tessier and W. Burleson. Reconfigurable Computing for Digital Signal Processing: A Survey. *Journal of VLSI Signal Processing Systems*, 28(1–2):7–27, 2001.

101. R. Hartenstein. A Decade of Reconfigurable Computing: A Visionary Retrospective. In *DATE '01: Proceedings of the conference on Design, automation and test in Europe*.

102. Sun Microsystems. *SpixTools: Introduction and User's Manual. http://research.sun.com/technical-reports/1993/abstract-6.html*.

103. Intel. *VTune. http://www.intel.com/cd/software/products/asmo-na/eng/vtune/239144.htm*.

104. Massimo Ravasi and Marco Mattavelli. High-level Algorithmic Complexity Evaluation for System Design. *Journal System Architectures*, 48(13–15):403–427, 2003.

105. K. Karuri, M. A. Al Faruque, S. Kraemer, R. Leupers, G. Ascheid and H. Meyr. Fine-grained Application Source Code Profiling for ASIP Design. In *DAC '05: Proceedings of the 42nd annual conference on Design automation*, pages 329–334, New York, NY, USA, 2005. ACM Press.

106. K. Karuri, C. Huben, R. Leupers, G. Ascheid and H. Meyr. Memory Access Micro-Profiling for ASIP Design. In *DELTA*, pages 255–262, 2006.

107. L. Pozzi, M. Vuleti and P. Ienne. Automatic Topology-Based Identification of Instruction-Set Extensions for Embedded Processors. In *DATE '02: Proceedings of the conference on Design, automation and test in Europe*, page 1138, Washington, DC, USA, 2002. IEEE Computer Society.

108. M. Arnold and H. Corporaal. Designing Domain-specific Processors. In *CODES '01: Proceedings of the ninth international symposium on Hardware/Software codesign*, pages 61–66, New York, NY, USA, 2001. ACM Press.

109. K. Atasu, L. Pozzi and P. Ienne. Automatic Application-specific Instruction-set Extensions Under Microarchitectural Constraints. In *DAC '03: Proceedings of the 40th conference on Design automation*, pages 256–261, New York, NY, USA, 2003. ACM Press.

110. P. Yu and T. Mitra. Scalable Custom Instructions Identification for Instruction-set Extensible Processors. In *CASES '04: Proceedings of the 2004 international conference on Compilers, architecture, and Synthesis for Embedded Systems*, pages 69–78, New York, NY, USA, 2004. ACM Press.

111. P. Bonzini and L. Pozzi. Polynomial-Time Subgraph Enumeration for Automated Instruction Set Extension. In *DATE: Proceedings of the Design, Automation and Test in Europe Conference*, 2007.

112. K. Atasu, G. Duendar and C. Oezturan. An Integer Linear Programming Approach for Identifying Instruction-set Extensions. In *CODES+ISSS '05: Proceedings of the 3rd IEEE/ACM/IFIP International Conference on Hardware/Software Codesign and System Synthesis*, pages 172–177, 2005.

113. R. Leupers, K. Karuri, S. Kraemer and M. Pandey. A Design Flow for Configurable Embedded Processors based on Optimized Instruction Set Extension Synthesis. In *Design, Automation & Test in Europe (DATE)*, Munich, Germany, March 2006.

114. K. Atasu, R. G. Dimond, O. Mencer, W. Luk, C. Oezturan and G. Duendar. Optimizing Instruction-set Extensible Processors under Data Bandwidth Constraints. In *Design, Automation & Test in Europe (DATE)*, Nice, France, April 2007.

115. P. Biswas, V. Choudhary, K. Atasu, L. Pozzi, P. Ienne and N. Dutt. Introduction of Local Memory Elements in Instruction Set Extensions. In *DAC '04: Proceedings of the 41st annual conference on Design automation*, 2004.

116. J. Cong, Y. Fan, G. Han, A. Jagannathan, G. Reinman and Z. Zhang. Instruction Set Extension with Shadow Registers for Configurable Processors. In *FPGA*, pages 99–106, 2005.

117. L. Pozzi and P. Ienne. Exploiting Pipelining to Relax Register-file Port Constraints of Instruction-Set Extensions. In *CASES '05: Proceedings of the 2005 International Conference on Compilers, Architectures and Synthesis for Embedded Systems*, pages 2–10, New York, NY, USA, 2005. ACM Press.

118. R. Jayaseelan, H. Liu and T. Mitra. Exploiting Forwarding to Improve Data Bandwidth of Instruction-set Extensions. In *DAC '06: Proceedings of the 43rd annual conference on Design automation*, pages 43–48, New York, NY, USA, 2006. ACM Press.

119. F. Sun, S. Ravi, A. Raghunathan and N. K. Jha. Synthesis of Custom Processors based on Extensible Platforms. In *ICCAD '02: Proceedings of the 2002 IEEE/ACM international conference on Computer-aided design*, pages 641–648, New York, NY, USA, 2002. ACM Press.

120. A. Peymandoust, L. Pozzi, P. Ienne and G. De Micheli. Automatic Instruction Set Extension and Utilization for Embedded Processors. In *14th IEEE International Conference on*

Application-Specific Systems, Architectures, and Processors (ASAP 2003), 24–26 June 2003, The Hague, The Netherlands. IEEE Computer Society, 2003.

121. N. Clark, J. Blome, M. Chu, S. Mahlke, S. Biles and K. Flautner. An Architecture Framework for Transparent Instruction Set Customization in Embedded Processors. In *ISCA '05: Proceedings of the 32nd Annual International Symposium on Computer Architecture*, pages 272–283, Washington, DC, USA, 2005. IEEE Computer Society.

122. G. Dittmann and P. Hurley. Instruction-set Synthesis for Reactive Real-Time Processors: An ILP Formulation. IBM Research Report, RZ 3611, IBM, 2005.

123. P. Yu and T. Mitra. Satisfying Real-time Constraints with Custom Instructions. In *CODES+ISSS '05: Proceedings of the 3rd IEEE/ACM/IFIP International Conference on Hardware/Software Codesign and System Synthesis*, pages 166–171, 2005.

124. E. M. Panainte, K. Bertels and S. Vassiliadis. Instruction Scheduling for Dynamic Hardware Configurations. In *DATE '05: Proceedings of the conference on Design, Automation and Test in Europe*, pages 100–105, Washington, DC, USA, 2005. IEEE Computer Society.

125. H. P. Huynh, J. E. Sim and T. Mitra. An Efficient Framework for Dynamic Reconfiguration of Instruction-set Customization. In *CASES '07: Proceedings of the 2007 International Conference on Compilers, Architecture, and Synthesis for Embedded Systems*, pages 135–144, New York, NY, USA, 2007. ACM Press.

126. M. Flynn. Some Computer Organizations and their Effectiveness. *IEEE Transactions on Computer*, C-21(9):948–960, 1972.

127. D. Sima, T. Fountain and P. Kacsuk. *Advanced Computer Architectures - A Design Space Approach.* Addison-Wesley, 1997.

128. D. A. Patterson and J. L. Hennessy. *Computer Organization and Design, 3rd Edition.* Morgan Kaufmann, 2004.

129. Gaisler Research. *http://www.gaisler.com/.*

130. R. Banakar, S. Steinke, B. Lee, M. Balakrishnan and P. Marwedel. Scratchpad Memory: A Design Alternative for Cache On-chip memory in Embedded Systems. In *Proc. of the 10th International Workshop on Hardware/Software Codesign, CODES, Estes Park (Colorado)*, May 2002.

131. G. Goossens. J. V. Praet, D. Lanneer, W. Geurts, A. Kifli, C. Liem and P. G. Paulin. Embedded Software in Real-Time Signal Processing Systems: Design Technologies. pages 433–451, 2002.

132. K. Okada, T. Ehara, H. Suzuki, K. Yanagida, K. Saito and N. Ichiura. A Digital Signal Processor Module Architecture and its Implementation using VLSIs. In *IEEE International Conference on Acoustics, Speech, and Signal Processing, ICASSP*, pages 378–381, 1984.

133. Actel. *http://www.actel.com.*

134. J. Rose, A. El Gamal and A. Sangiovanni-Vincentelli. Architecture of Field-Programmable Gate Arrays. *Proceedings of the IEEE*, 81(7):1013–1029, 1993.

135. P. M. Heysters, G. J. M. Smit and E. Molenkamp. Energy-Efficiency of the Montium Reconfigurable Tile Processor. In *Proceedings of the International Conference on Engineering of Reconfigurable Systems and Algorithms (ERSA'04), Las Vegas, USA*, pages 38–44, June 2004.

136. A. Hoffmann, T. Kogel, A. Nohl, G. Braun, O. Schliebusch, O. Wahlen, A. Wieferink and H. Meyr. A Novel Methodology for the Design of Application Specific Instruction-Set Processor Using a Machine Description Language. In *IEEE Transactions on Computer-Aided Design of Integrated Cicuits and Systems (TCAD). Vol. 20 no.11*, pages 1338–1354. IEEE, 2001.

137. A. Chattopadhyay, X. Chen, H. Ishebabi, R. Leupers, G. Ascheid and H. Meyr. High-Level Modelling and Exploration of Coarse-Grained Re-Configurable Architectures. In *Proceedings of the Conference on Design, Automation and Test in Europe*, Munich, Germany, March 2008.

138. J. O. Filho, S. Masekowsky, T. Schweizer and W. Rosenstiel. An Architecture Description Language for Coarse-Grained Reconfigurable Arrays. In *Workshop on Architecture Specific Processors (WASP)*, 2007.

139. Synopsys. *Design Compiler. http://www.synopsys.com/products/logic/design_compiler.html.*

140. P. Biswas, S. Banerjee, N. Dutt, L. Pozzi and P. Ienne. ISEGEN: an Iterative Improvement-based ISE Generation Technique for Fast Customization of Processors. *IEEE Transactions on VLSI Systems*, 14(7), 2006.

141. H. Blume, S. Kannengiesser and T.G. Noll. Image Quality Enhancement for MRT Images. In *Proceedings of the ProRISC Workshop*, Veldhoven, Netherlands, 2003.

142. L. Bauer, M. Shafique, S. Kramer and J. Henkel. RISPP: Rotating Instruction Set Processing Platform. In *DAC '07: Proceedings of the 44th annual conference on Design automation*, pages 791–796, New York, NY, USA, 2007. ACM Press.

143. F.T. Leighton. A graph coloring algorithm for large scheduling problems. *Journal of Research of the National Bureau of Standards*, 84:489–506, 1979.

144. JJ. Ceng, M. Hohenauer, R. Leupers, G. Ascheid, H. Meyr and G. Braun. C Compiler Retargeting Based on Instruction Semantics Models. In *Proceedings of the Conference on Design, Automation & Test in Europe (DATE)*, Munich, Germany, March 2005.

145. M. Hohenauer, C. Schumacher, R. Leupers, G. Ascheid, H. Meyr and H. van Someren. Retargetable Code Optimization with SIMD Instructions. In *CODES+ISSS '06: Proceedings of the 4th International Conference on Hardware/Software Codesign and System Synthesis*, pages 148–153, New York, NY, USA, 2006. ACM.

146. O. Wahlen, M. Hohenauer, R. Leupers and H. Meyr. Instruction Scheduler Generation for Retargetable Compilation. In *IEEE Design & Test of Computers*. IEEE, Jan/Febr 2003.

147. Cadence. *Encounter Digital IC Design Platform. http://www.cadence.com/products/digital_ic/index.aspx.*

148. Magma Design Automation. *Blast Fusion. http://www.magma-da.com.*

149. O. Schliebusch, A. Hoffmann, A. Nohl, G. Braun and H. Meyr. Architecture Implementation Using the Machine Description Language LISA. In *Proc. of the ASPDAC/VLSI Design - Bangalore, India*, 2002.

150. O. Schliebusch, A. Chattopadhyay, D. Kammler, R. Leupers, G. Ascheid and H. Meyr. A Framework for Automated and Optimized ASIP Implementation Supporting Multiple Hardware Description Languages. In *Proc. of the ASPDAC - Shanghai, China*, 2005.

151. P. Mishra, A. Kejariwal, and N. Dutt. Synthesis-driven Exploration of Pipelined Embedded Processors. In *Int. Conf. on VLSI Design*, 2004.

152. S. Basu and R. Moona. High Level Synthesis from Sim-nML Processor Models. In *VLSI Design*, 2003.

153. E. M. Witte, A. Chattopadhyay, O. Schliebusch, D. Kammler, G. Ascheid, R. Leupers and H. Meyr. Applying Resource Sharing Algorithms to ADL-driven Automatic ASIP Implementation. In *ICCD '05: Proceedings of the 2005 International Conference on Computer Design*, pages 193–199, Washington, DC, USA, 2005. IEEE Computer Society.

154. O. Schliebusch, A. Chattopadhyay, E. M. Witte, D. Kammler, G. Ascheid, R. Leupers and H. Meyr. Optimization Techniques for ADL-Driven RTL Processor Synthesis. In *RSP '05: Proceedings of the 16th IEEE International Workshop on Rapid System Prototyping (RSP'05)*, pages 165–171, Washington, DC, USA, 2005. IEEE Computer Society.

155. A. Chattopadhyay, B. Geukes, D. Kammler, E. M. Witte, O. Schliebusch, H. Ishebabi, R. Leupers, G. Ascheid and H. Meyr. Automatic ADL-Based Operand Isolation for Embedded Processors. In *DATE '06: Proceedings of the Conference on Design, Automation and Test in Europe*, pages 600–605, 3001 Leuven, Belgium, Belgium, 2006. European Design and Automation Association.

156. A. Chattopadhyay, D. Kammler, E. M. Witte, O. Schliebusch, H. Ishebabi, B. Geukes, R. Leupers, G. Ascheid and H. Meyr. Automatic Low Power Optimizations During ADL-Driven ASIP Design. In *IEEE International Symposium on VLSI Design, Automation and Test (VLSI-DAT)*, Hsinchu, Taiwan, April 2006.

157. E. M. Witte. Analysis and Implementation of Resource Sharing Optimizations for RTL Processor Synthesis. Master's thesis, Integrated Signal Processing Systems, RWTH Aachen, 2004. Advisor: O. Schliebusch, A. Chattopadhyay, G. Ascheid, H. Meyr.

158. S. Sirowy, Y. Wu, S. Lonardi and F. Vahid. Two-level Microprocessor-Accelerator Partitioning. In *DATE '07: Proceedings of the conference on Design, Automation and Test in Europe*, pages 313–318, New York, NY, USA, 2007. ACM Press.

159. A. Chattopadhyay, D. Kammler, D. Zhang, G. Ascheid, R. Leupers and H. Meyr. Specification-Driven Exploration and Implementation of Partially Re-configurable Processors. In *GSPx*, Santa Clara, California, USA, October–November 2006.

160. E. Haseloff. Metastable Response in 5-V Logic Circuits. *http://focus.ti.com/lit/an/sdya006/sdya006.pdf*, 1997.

161. D. Kammler. Design of a Hardware Debug Interface for Processors Described with LISA. Master's thesis, Integrated Signal Processing Systems, RWTH Aachen, 2004. Advisor: O. Schliebusch, H. Meyr.

162. R. Weinstein. The Flancter. *Xilinx Xcell*, 2000.

163. J. Y. Lin, D. Chen and J. Cong. Optimal Simultaneous Mapping and Clustering for FPGA Delay Optimization. In *DAC '06: Proceedings of the 43rd annual conference on Design automation*, pages 472–477, New York, NY, USA, 2006. ACM Press.

164. G. Dupenloup, T. Lemeunier and R. Mayr. Transistor Abstraction for the Functional Verification of FPGAs. In *DAC '06: Proceedings of the 43rd annual conference on Design automation*, pages 1069–1072, New York, NY, USA, 2006. ACM Press.

165. M. Gansen, F. Richter, O. Weiss and T. G. Noll. A Datapath Generator for Full-Custom Macros of Iterative Logic Arrays. In *ASAP '97: Proceedings of the IEEE International Conference on Application-Specific Systems, Architectures and Processors*, page 438, Washington, DC, USA, 1997. IEEE Computer Society.

166. K. Eguro and S. Hauck. Issues and Approaches to Coarse-grain Reconfigurable Architecture Development. In *11th Annual IEEE Symposium on Field-Programmable Custom Computing Machines (FCCM)*, pages 111–120, 2003.

167. V. Betz, J. Rose and A. Marquardt, editor. *Architecture and CAD for Deep-Submicron FPGAs*. Kluwer Academic Publishers, Norwell, MA, USA, 1999.

168. E. L. Lawler, K. N. Levitt and J. Turner. Module Clustering to Minimize Delay in Digital Networks. *IEEE Transactions on Computers*, C-18(1):47–57, 1969.

169. R. Murgai, R.K. Brayton and A. Sangiovanni-Vincentelli. On Clustering for Minimum Delay/Area. In *ICCAD '91: Proceedings of the 1991 IEEE/ACM International Conference on Computer-Aided Design*, pages 6–9, 1991.

170. H. Yang and D. F. Wong. Edge-map: Optimal Performance Driven Technology Mapping for Iterative LUT-based FPGA Designs. In *ICCAD '94: Proceedings of the 1994 IEEE/ACM International Conference on Computer-aided Design*, pages 150–155, Los Alamitos, CA, USA, 1994. IEEE Computer Society Press.

171. R. Francis, J. Rose and Z. Vranesic. Chortle-crf: Fast Technology Mapping for Lookup Table-based FPGAs. In *DAC '91: Proceedings of the 28th Conference on ACM/IEEE Design Automation*, pages 227–233, New York, NY, USA, 1991. ACM Press.

172. J. H. Anderson and F. N. Najm. Power-aware Technology Mapping for LUT-based FPGAs. In *Proceedings of the 2002 IEEE International Conference on Field-Programmable Technology*, pages 211–218, 2002.

173. J. Cong and Y. Ding. An Optimal Technology Mapping Algorithm for Delay Optimization in Lookup-table based FPGA Designs. In *ICCAD '92: Proceedings of the 1992 IEEE/ACM International Conference on Computer-Aided Design*, pages 48–53, Los Alamitos, CA, USA, 1992. IEEE Computer Society Press.

174. R. Rajaraman and D. F. Wong. Optimal Clustering for Delay Minimization. In *DAC '93: Proceedings of the 30th International Conference on Design Automation*, pages 309–314, New York, NY, USA, 1993. ACM Press.

175. A. Sharma, C. Ebeling and S. Hauck. Architecture Adaptive Routability-Driven Placement for FPGAs. In *FPGA '05: Proceedings of the 2005 ACM/SIGDA 13th International Symposium on Field-programmable Gate Arrays*, pages 266–266, New York, NY, USA, 2005. ACM Press.

176. V. Betz and J. Rose. VPR: A new Packing, Placement and Routing Tool for FPGA Research. In W. Luk, P. Y. K. Cheung and M. Glesner, editor, *Field-Programmable Logic and Applications*, pages 213–222. Springer-Verlag, Berlin, 1997.

177. S. Kirkpatrick and C. D. Gelatt and M. P. Vecchi. Optimization by Simulated Annealing. *Science*, 220, 4598:671–680, 1983.

178. J. Frankle. Iterative and Adaptive Slack Allocation for Performance-Driven Layout and FPGA Routing. In *DAC '92: Proceedings of the 29th ACM/IEEE Conference on Design Automation*, pages 536–542, Los Alamitos, CA, USA, 1992. IEEE Computer Society Press.

179. L. McMurchie and C. Ebeling. PathFinder: A Negotiation-based Performance-driven Router for FPGAs. In *FPGA '95: Proceedings of the 1995 ACM third International Symposium on Field-Programmable Gate Arrays*, pages 111–117, New York, NY, USA, 1995. ACM Press.

180. Electrical and Computer Engineering Department, University of Toronto, Canada. *The FPGA Place-and-Route Challenge – http://www.eecg.toronto.edu/~vaughn/challenge/challenge.html*.

181. N. Bansal, S. Gupta, N. Dutt and A. Nicolau. Analysis of the Performance of Coarse-Grain Reconfigurable Architectures with Different Processing Element Configurations. In *Workshop on Architecture Specific Processors (WASP)*, 2003.

182. N. Bansal, S. Gupta, N. Dutt, A. Nicolau and R. Gupta. Network Topology Exploration of Mesh-based Coarse-grain Reconfigurable Architectures. Technical report.

183. K. Keutzer. DAGON: Technology Binding and Local Optimization by DAG Matching. In *DAC '87: Proceedings of the 24th ACM/IEEE conference on Design automation*, pages 341–347, New York, NY, USA, 1987. ACM Press.

184. G. Dimitroulakos, M. D. Galanis and C. E. Goutis. Design Space Exploration of an Optimized Compiler Approach for a Generic Reconfigurable Array Architecture. *Journal of Supercomputing*, 40(2):127–157, 2007.

185. Y. Guo, G. J.M. Smit, H. Broersma and P. M. Heysters. A Graph Covering Algorithm for a Coarse Grain Reconfigurable System. In *LCTES '03: Proceedings of the 2003 ACM SIGPLAN Conference on Language, Compiler and Tool for Embedded Systems*, pages 199–208, New York, NY, USA, 2003. ACM Press.

186. A. Ye, J. Rose and L. David. Architecture of Datapath-oriented coarse-grain logic and routing for FPGAs. In *Proceedings of the IEEE 2003 Custom Integrated Circuits Conference*, pages 61–64, 2003.

187. S. Sahni and A. Bhatt. The Complexity of Design Automation Problems. In *DAC '80: Proceedings of the 17th Conference on Design Automation*, pages 402–411, New York, NY, USA, 1980. ACM Press.

188. A. Chattopadhyay, W. Ahmed, K. Karuri, D. Kammler, R. Leupers, G. Ascheid and H. Meyr. Design Space Exploration of Partially Re-Configurable Embedded Processors. In *Proceedings of the Conference on Design, Automation and Test in Europe*, Nice, France, April 2007.

189. A. Chattopadhyay, H. Ishebabi, X. Chen, Z. Rakosi, K. Karuri, D. Kammler, R. Leupers, G. Ascheid and H. Meyr. Prefabrication and Postfabrication Architecture Exploration for Partially Reconfigurable VLIW Processors. *ACM Transactions on Embedded Computing Systems*, 7(4), 2008.

190. Synopsys. *http://www.synopsys.com*.

191. Xilinx. *Virtex-II pro. http://www.xilinx.com/products/silicon_solutions/fpgas/*.

192. Nexperia. *TriMedia Programmable Processor. http://www.nxp.com/*.

193. J. Becker, T. Piontek and M. Glesner. DReAM: A Dynamically Reconfigurable Architecture for Future Mobile Communications Applications. In *FPL '00: Proceedings of the The Roadmap to Reconfigurable Computing, 10th International Workshop on Field-Programmable Logic and Applications*, pages 312–321, London, UK, 2000. Springer-Verlag.

194. Y. Lin, H. Lee, M. Woh, Y. Harel, S. Mahlke, T. Mudge, C. Chakrabarti and K. Flautner. SODA: A High-Performance DSP Architecture for Software-Defined Radio. *IEEE Micro*, 27(1):114–123, 2007.

195. D. Novo Bruna, B. Bougard, P. Raghavan, T. Schuster, K. Hong-Seok, Y. Ho, and L. Van der Perre. Energy-Performance Exploration of a CGA-Based SDR Processor. In *Software Defined Radio Technical Conference and Product Exposition*, 2006.

196. A.S.Y. Poon. An Energy-Efficient Reconfigurable Baseband Processor for Flexible Radios. In *SIPS '06 : IEEE Workshop on Signal Processing Systems Design and Implementation*, pages 393–398, 2006.

197. H. Holma and A. Toskala. *WCDMA for UMTS: Radio Access for Third Generation Mobile Communications*. Wiley Technology Publishing, 2000.

198. CoWare/Signal Processing Designer. *http://www.coware.com*.

199. H. Ishebabi, D. Kammler, G. Ascheid, H. Meyr, M. Nicola, G. Masera, M. Zamboni. FFT Processor: A Case Study in ASIP Development. In *IST : Mobile and Wireless Communications Summit*, Dresden, Germany, June 2005.

200. L. J. Garin. The "Shaping Correlator", Novel Multipath Mitigation Technique Applicable to GALILEO BOC(1,1) Modulation Waveforms in High Volume Markets. In *Proceedings of ENC-GNSS*, 2005.

201. G. Kappen and T. G. Noll. Application Specific Instruction Processor Based Implementation of a GNSS Receiver on an FPGA. In *DATE '06: Proceedings of the conference on Design, Automation and Test in Europe*, pages 58–63, 3001 Leuven, Belgium, Belgium, 2006. European Design and Automation Association.

202. K. Karuri, A. Chattopadhyay, M. Hohenauer, R. Leupers, G. Ascheid and H. Meyr. Increasing Data-Bandwidth to Instruction-Set Extensions Through Register Clustering. In *IEEE/ACM International Conference on Computer-Aided Design (ICCAD)*, 2007.

203. Altera. *NIOS II. http://www.altera.com/*.

Appendix A
LISA Grammar for Coarse-Grained FPGA

In the following, the modified LISA grammar for parsing the coarse-grained FPGA's structural description is presented. The detailed LISA grammar is available at [10].

fpga : *fpga fpga_component* { }
| *fpga_component* { };

fpga_component
: *t_ELEMENT* '{' *element_part_list* '}' { }
| *t_TOPOLOGY* '{' *cluster_list* '}' { }
| *t_CONNECTIVITY* '{' *rule_list* '}' { };

element_part_list
: *element_part_list t_ELEMENT t_IDENTIFIER* '{' *port attributes behavior* '}' { }
| *t_ELEMENT t_IDENTIFIER* '{' *port attributes behavior* '}' { };

port
: *t_PORT* '{' *port_declaration_list* '}' { };

port_declaration_list
: *port_declaration_list port_declaration* { }
| *port_declaration* { };

port_declaration
: *t_IN sign_declaration* '<' *t_NUMBER* '>' *identifier_list* ';' { }
| *t_OUT sign_declaration* '<' *t_NUMBER* '>' *identifier_list* ';' { };

sign_declaration
: *t_SIGNED* { }
| *t_UNSIGNED* { };

behavior
: *t_BEHAVIOR* '{' *fpga_c_code* '}' { };

cluster_list
: *cluster_list cluster* { }
| *cluster* { };

    cluster
: *t_CLUSTER t_IDENTIFIER* '{' *cluster_port layout attributes* '}' {};

    cluster_port
: *t_PORT* '{' *cluster_port_declaration_list* '}' {};

    cluster_port_declaration_list
: *cluster_port_declaration_list cluster_port_declaration* {}
| *cluster_port_declaration* {};

    cluster_port_declaration
: *t_IN sign_declaration* '<' *t_NUMBER* '>' *cluster_port_identifier_list* ';' {}
| *t_IN sign_declaration* '<' *t_NUMBER* '>' *array_declaration cluster_port_identifier_list* ';' {}
| *t_OUT sign_declaration* '<' *t_NUMBER* '>' *cluster_port_identifier_list* ';' {}
| *t_OUT sign_declaration* '<' *t_NUMBER* '>' *array_declaration cluster_port_identifier_list* ';' {};

    cluster_port_identifier_list
: *cluster_port_identifier_list* ' ' *t_IDENTIFIER* {}
| *t_IDENTIFIER* {};

    layout
: *t_LAYOUT* '{' *layout_list* '}' {};

    layout_list
: *layout_list layout_decl* {}
| *layout_decl* {};

    layout_decl
: *t_ROW t_IDENTIFIER* '=' '{' *entity_identifier_list* '}' ';' {};

    entity_identifier_list
: *entity_identifier_list* ',' *t_IDENTIFIER* {}
| *t_IDENTIFIER* {};

    rule_list
: *t_RULE t_IDENTIFIER* '{' *style_list basic_list* '}' *rule_list* {}
| *t_RULE t_IDENTIFIER* '{' *style_list basic_list* '}' {};

    style_list
: *style_list style_declaration* {}
| *style_declaration* {};

    style_declaration
: *t_STYLE* '(' *style_type* ',' *t_NUMBER* ')' ';' {}
| t_STYLE '(' *style_type* ')' ';' {};

    style_type
: *t_POINT2POINT* {}
| *t_MESH* {}
| *t_NEARESTNEIGHBOUR* {}
| *t_ROWWISE* {}
| *t_COLUMNWISE* {};

basic_list
: basic_declaration basic_list {}
| basic_declaration {};

basic_declaration
: t_BASIC '(' t_IDENTIFIER ',' t_IDENTIFIER ')' '{' basic_body '}' {};

basic_body
: basic_statement basic_body {}
| basic_statement {};

basic_statement
: src_connector t_TO dst_connector_list ';' {}
| src_connector array_access t_TO dst_connector_list ';' {}
| src_connector t_TO dst_connector array_access ';' {}
| src_connector t_TO dst_connector subscription ';' {};

subscription
: t_OPEN_SQ_BRACKET t_NUMBER t_CLOSE_SQ_BRACKET {};

array_access
: t_OPEN_SQ_BRACKET t_NUMBER t_DOUB_DOT t_NUMBER t_CLOSE_SQ_BRACKET {};

src_connector
: t_IDENTIFIER '.' t_IDENTIFIER {};

dst_connector_list
: dst_connector t_OR dst_connector_list {}
| dst_connector {};

dst_connector
: t_IDENTIFIER '.' t_IDENTIFIER {};

attributes
: t_ATTRIBUTES '{' registered bypass '}' {}
| t_ATTRIBUTES '{' bypass registered '}' {};

bypass
: t_BYPASS '(' identifier_list ')' ';' {}
| t_BYPASS '(' t_NONE ')' ';' {};

registered
: t_REGISTER '(' identifier_list ')' ';' {}
| t_REGISTER '(' t_NONE ')' ';' {};

array_declaration
: t_OPEN_SQ_BRACKET t_NUMBER t_DOUB_DOT t_NUMBER t_CLOSE_SQ_BRACKET {};

identifier_list
: identifier_list ',' t_IDENTIFIER {}
| t_IDENTIFIER {};

Printed in the United States
135037LV00003B/65/P